METHODS IN MOLECULAR BIOLOGY

Series Editor
John M. Walker
School of Life Sciences
University of Hertfordshire
Hatfield, Hertfordshire, AL10 9AB, UK

For further volumes:
http://www.springer.com/series/7651

Mitochondrial Regulation

Methods and Protocols

Edited by

Carlos M. Palmeira

*Department of Life Sciences and Center for Neurosciences and Cell Biology,
University of Coimbra, Coimbra, Portugal*

Anabela P. Rolo

*Department of Biology, University of Aveiro, Aveiro, Portugal;
Center for Neurosciences and Cell Biology, University of Coimbra, Coimbra, Portugal*

 Humana Press

Editors
Carlos M. Palmeira
Department of Life Sciences and Center for
 Neurosciences and Cell Biology
University of Coimbra
Coimbra, Portugal

Anabela P. Rolo
Department of Biology
University of Aveiro
Aveiro, Portugal

Center for Neurosciences and Cell Biology
University of Coimbra
Coimbra, Portugal

ISSN 1064-3745   ISSN 1940-6029 (electronic)
ISBN 978-1-4939-1874-4   ISBN 978-1-4939-1875-1 (eBook)
DOI 10.1007/978-1-4939-1875-1
Springer New York Heidelberg Dordrecht London

Library of Congress Control Number: 2014949856

Printed on acid-free paper

Humana Press is a brand of Springer
Springer is part of Springer Science+Business Media (www.springer.com)

Preface

This book attempts to show different pathways that converge into regulation of mitochondrial function. The book will integrate mitochondria with other cellular components, discussing how these interactions influence the dynamics of mitochondrial structure and biogenesis.

This book is intended for the use of advanced undergraduates, graduates, postgraduates, and beginning researchers in the areas of molecular and cellular biology, biochemistry, and bioenergetics. Each section is prefaced with a short treatise introducing the fundamental principles for that section followed by chapters describing the practical principles and assays designed to derive quantitative assessment of each set of parameters that reflect different aspects of mitochondrial regulation.

Coimbra, Portugal   *Carlos M. Palmeira*
Aveiro, Portugal   *Anabela P. Rolo*

Contents

Contributors

ORNELLA DE BARI • *Department of Biomedical Sciences and Human Oncology, Clinica Medica "A. Murri", "Aldo Moro" University Medical School, Bari, Italy; Division of Gastroenterology and Hepatology, Department of Internal Medicine, Saint Louis University School of Medicine, St. Louis, MO, USA*

LEONILDE BONFRATE • *Department of Biomedical Sciences and Human Oncology, Clinica Medica "A. Murri", "Aldo Moro" University Medical School, Bari, Italy*

PEDRO M. BORRALHO • *Faculty of Pharmacy, Research Institute for Medicines (iMed. UL isboa), Universidade de Lisboa, Lisboa, Portugal*

SILVIA CAMPELLO • *Department of Experimental Neuroscience, IRCCS Fondazione Santa Lucia, Rome, Italy*

AUDREY M. CARROLL • *School of Biochemistry and Immunology, Trinity Biomedical Sciences Institute, Trinity College Dublin, Dublin, Ireland*

LUIGI CASTORANI • *Department of Biomedical Sciences and Human Oncology, Clinica Medica "A. Murri", "Aldo Moro" University Medical School, Bari, Italy*

FEDERICA CIOFFI • *Dipartimento di Scienze e Tecnologie, Università degli Studi del Sannio, Benevento, Italy*

KIERAN J. CLARKE • *School of Biochemistry and Immunology, Trinity Biomedical Sciences Institute, Trinity College Dublin, Dublin, Ireland*

MERYL D. COLTON • *Nicholas School of the Environment, Duke University, Durham, NC, USA*

MAURO CORRADO • *Department of Experimental Neuroscience, IRCCS Fondazione Santa Lucia, Roma, Italy; Dulbecco-Telethon Institute, Venetian Institute of Molecular Medicine, Padova, Italy*

TERESA CUNHA-OLIVEIRA • *CNC—Center for Neuroscience and Cell Biology, Largo Marqués de Pombal, University of Coimbra, Coimbra, Portugal*

NIKA N. DANIAL • *Department of Cancer Biology, Dana-Farber Cancer Institute, Boston, MA, USA; Department of Cell Biology, Harvard Medical School, Boston, MA, USA*

KAYLYN E. GERM • *The Institute of Environmental and Human Health, Texas Tech University, Lubbock, TX, USA*

ALFREDO GIMÉNEZ-CASSINA • *Department of Cancer Biology, Dana-Farber Cancer Institute, Boston, MA, USA; Department of Cell Biology, Harvard Medical School, Boston, MA, USA*

FERNANDO GOGLIA • *Dipartimento di Scienze e Tecnologie, Università degli Studi del Sannio, Benevento, Italy*

ANA P. GOMES • *Department of Genetics, Harvard Medical School, Boston, MA, USA*

IGNAZIO GRATTAGLIANO • *Italian College of General Practitioners, Florence, Bari, Italy*

J. TIMOTHY GREENAMYRE • *Department of Neurology, Pittsburgh Institute for Neurodegenerative Diseases, University of Pittsburgh, Pittsburgh, PA, USA*

EVAN H. HOWLETT • *Department of Neurology, Pittsburgh Institute for Neurodegenerative Diseases, University of Pittsburgh, Pittsburgh, PA, USA*

AGNIESZKA KARKUCINSKA-WIECKOWSKA • *Department of Pathology, The Children's Memorial Health Institute, Warsaw, Poland*

WEI LI • *Weill Cornell Medical College, New York, NY, USA*

ASSUNTA LOMBARDI • *Dipartimento delle Scienze Biologiche, Sezione Fisiologia, Università degli Studi di Napoli "Federico II", Naples, Italy*

MICHELE LORUSSO • *Department of Biomedical Sciences and Human Oncology, Clinica Medica "A. Murri", "Aldo Moro" University Medical School, Bari, Italy*

FRANCESCA R. MARIOTTI • *Department of Experimental Neuroscience, IRCCS Fondazione Santa Lucia, Rome, Italy*

GREG D. MAYER • *The Institute of Environmental and Human Health, Texas Tech University, Lubbock, TX, USA*

JOEL N. MEYER • *Nicholas School of the Environment, Duke University, Durham, NC, USA*

GEMMA O'BRIEN • *School of Biochemistry and Immunology, Trinity Biomedical Sciences Institute, Trinity College Dublin, Dublin, Ireland*

PAULO J. OLIVEIRA • *CNC—Center for Neuroscience and Cell Biology, Largo Marquês de Pombal, University of Coimbra, Coimbra, Portugal*

CARLOS M. PALMEIRA • *Center for Neurosciences and Cell Biology, University of Coimbra, Coimbra, Portugal; Department of Life Sciences, University of Coimbra, Coimbra, Portugal*

SIMONE PATERGNANI • *Department of Morphology, Surgery and Experimental Medicine, Section of Pathology, Oncology and Experimental Biology, Laboratory for Technologies of Advanced Therapies (LTTA), University of Ferrara, Ferrara, Italy*

PAOLO PINTON • *Department of Morphology, Surgery and Experimental Medicine, Section of Pathology, Oncology and Experimental Biology, Laboratory for Technologies of Advanced Therapies (LTTA), University of Ferrara, Ferrara, Italy*

RICHARD K. PORTER • *School of Biochemistry and Immunology, Trinity Biomedical Sciences Institute, Trinity College Dublin, Dublin, Ireland*

PIERO PORTINCASA • *Department of Biomedical Sciences and Human Oncology, Clinica Medica "A. Murri", "Aldo Moro" University Medical School, Bari, Italy*

MACIEJ PRONICKI • *Department of Pathology, The Children's Memorial Health Institute, Warsaw, Poland*

FABRICE RAPPAPORT • *UMR 7141 CNRS-UPMC, Institut de Biologie Physico-Chimique, Paris, France*

SOFIA M. RIBEIRO • *Department of Cancer Biology, Dana-Farber Cancer Institute, Boston, MA, USA; Ph.D. Programme in Experimental Biology and Biomedicine (PDBEB), CNC—Center for Neuroscience and Cell Biology, University of Coimbra, Coimbra, Portugal; Institute for Interdisciplinary Research (IIIUC)*

CECÍLIA M.P. RODRIGUES • *Faculty of Pharmacy, Research Institute for Medicines and (iMed. UL isboa), Universidade de Lisboa, Lisboa, Portugal*

ANABELA P. ROLO • *Center for Neurosciences and Cell Biology, University of Coimbra, Coimbra, Portugal; Department of Biology, University of Aveiro, Aveiro, Portugal*

JOHN P. ROONEY • *Nicholas School of the Environment, Duke University, Durham, NC, USA*

IAN T. RYDE • *Nicholas School of the Environment, Duke University, Durham, NC, USA*

LAURIE H. SANDERS • *Department of Neurology, Pittsburgh Institute for Neurodegenerative Diseases, University of Pittsburgh, Pittsburgh, PA, USA*

VILMA SARDÃO • *CNC—Center for Neuroscience and Cell Biology, Largo Marquês de Pombal, University of Coimbra, Coimbra, Portugal*

ANTHONY A. SAUVE • *Weill Cornell Medical College, New York, NY, USA*

TERESA SERAFIM • *CNC—Center for Neuroscience and Cell Biology, Largo Marquês de Pombal, University of Coimbra, Coimbra, Portugal*

ELENA SILVESTRI • *Dipartimento di Scienze e Tecnologie, Università degli Studi del Sannio, Benevento, Italy*

DAVID A. SINCLAIR • *Department of Genetics, Harvard Medical School, Boston, MA, USA*

CLIFFORD J. STEER • *Departments of Medicine, and Genetics, Cell Biology and Development, University of Minnesota Medical School, Minneapolis, MN, USA*

JOÃO S. TEODORO • *Center for Neurosciences and Cell Biology, University of Coimbra, Coimbra, Portugal*

IGNACIO VEGA-NAREDO • *CNC—Center for Neuroscience and Cell Biology, Largo Marquês de Pombal, University of Coimbra, Coimbra, Portugal*

MARIUSZ R. WIECKOWSKI • *Laboratory of Bioenergetics and Biomembranes, Department of Biochemistry, Nencki Institute of Experimental Biology, Warsaw, Poland*

LECH WOJTCZAK • *Nencki Institute of Experimental Biology, Warsaw, Poland*

Chapter 1

Isolation of Crude Mitochondrial Fraction from Cells

Mariusz R. Wieckowski and Lech Wojtczak

Abstract

Mitochondria are intracellular organelles where most fundamental processes of energy transformation within the cell are located. They also play important roles in programmed cell death (apoptosis), free radical formation, and signal transduction. In addition, mitochondria host genes encoding several important proteins. Studying isolated mitochondria is therefore crucial for better understanding cell functioning. The present article describes a relatively simple and handy procedure for isolation of crude mitochondrial fraction from cultivated mammalian and human cells. It consists of mechanical homogenization and fractionating centrifugation. Assays of checking mitochondrial competence by measuring membrane potential formation and coupled respiration are also presented.

Key words Mitochondria, Isolation, Cell cultures, Oxygen consumption, Membrane potential, Oxidative phosphorylation

1 Introduction

Methods of isolation of mitochondria from organs and solid tissues like liver, heart, brain, and skeletal muscles have been published already at late 1940s of the last century (*see* for example ref. [1]). They are based on disintegration of the tissue by homogenization or mincing with rotating knives (Polytron), sometimes with the help of proteolytic enzymes (trypsin, nagarse), followed by fractionating centrifugation. Isolation of mitochondrial from natural cell suspensions, as lymphocytes, or from cell cultures is complicated by the fact that some kinds of cells are more resistant to mechanic disruption than cells of solid tissues. To overcome this problem, homogenization has often been preceded, or even replaced, by disruption of the plasma membrane with the use of heparin [2] or digitonin [3], by nitrogen cavitation [4] or shearing cell suspensions through narrow needles. Nevertheless, homogenization remains the most universal procedure. It is mostly performed using a glass Potter–Elvehjem homogenizer with a motor-driven Teflon pestle (glass/Teflon homogenizer).

Carlos M. Palmeira and Anabela P. Rolo (eds.), *Mitochondrial Regulation*, Methods in Molecular Biology, vol. 1241, DOI 10.1007/978-1-4939-1875-1_1, © Springer Science+Business Media New York 2015

Dounce-type glass homogenizers with manually driven glass pestles are also used occasionally.

No matter how cells are disrupted, mitochondria have to be separated from the cytosol and other particulate organelles (nuclei, endoplasmic reticulum, etc.). The most common and simplest method to this aim is fractionating centrifugation. Further purification can be achieved by density-gradient centrifugation. A protocol for obtaining a highly purified mitochondrial fraction suitable for proteomic studies using Percol gradient centrifugation has been elaborated in this laboratory [5]. An alternative sophisticated procedure using magnetic microbeads conjugated to an antibody against one of mitochondrial outer membrane proteins (TOM22) has been proposed by Hornig-Do et al. [6]. Some new protocols are based on commercially available mitochondrial isolation kits. It has also to be mentioned that for certain purposes mitochondria need not to be extracted out of the cell and isolated as a separate fraction. We [7] have shown for example that certain properties of mitochondria like coupled respiration and membrane potential formation can be studied within cells after their controlled permeabilization with digitonin (the so-called "mitochondria in situ").

A comprehensive description of modern methods for isolation of mitochondria from a broad variety of biological material, including cultured cells, has been published by Palotti and Lenaz [8]. The present article describes a simple procedure to isolate a crude fraction of functionally competent mitochondria from cell cultures (either in suspension or plated) using homogenization and fractionating centrifugation.

2   Materials

2.1   Isolation of Crude Mitochondria Fraction from Cell Lines

2.1.1   Solutions and Chemicals

1. Homogenization medium: 75 mM sucrose (*see* **Note 1**), 225 mM mannitol, 0.1 mM EGTA, 30 mM Tris–HCl pH 7.4. Check pH with pH-meter and adjust if necessary. Store at 4 °C (*see* **Note 2**).

2. Mitochondria isolation medium: 75 mM sucrose (*see* **Note 1**), 225 mM mannitol, 5 mM Tris–HCl, pH 7.4. Check pH with pH-meter and adjust if necessary. Store at 4 °C (*see* **Note 2**).

3. Phosphatase inhibitor cocktail 3.

4. Protease inhibitor cocktail.

2.1.2   Equipment and Accessories

1. Stirrer motor with electronic speed controller.

2. Motor-driven tightly fitting glass/Teflon Potter–Elvehjem homogenizer.

3. Loose fitting glass/Teflon Potter–Elvehjem homogenizer.

2.2 Cell Culture

2.2.1 Materials and Accessories

1. Cells: cell line of interest, e.g., HeLa, MEF or NHDF, neonatal dermal fibroblasts growing in monolayer or cells growing in suspension like Jurkat (T-cell leukemia cell line) or rat hepatoma AS-30D cell line.

2. Cell culture dishes 10 cm diameter.

3. Cell scrapers.

2.2.2 Solutions

1. Dulbecco's Modified Eagle's Medium (DMEM) supplemented with 10 % fetal bovine serum (heat inactivated FBS), 2 mM L-glutamine and 1.2 % antibiotic: penicillin/streptomycin (Penicillin–Streptomycin Solution).

2. Dulbecco's Phosphate Buffered Saline (D-PBS), without Ca^{2+} and Mg^{2+}.

3. Dulbecco's Phosphate Buffered Saline (D-PBS), with Ca^{2+} and Mg^{2+}.

2.3 Determination of Protein Concentration

2.3.1 Equipment and Accessories

1. Spectrophotometer UV/Vis.

2. Bio-Rad protein assay.

3. Acryl cuvettes $10 \times 10 \times 48$ mm for spectrophotometer.

4. Bovine serum albumin standard solution, 1 mg/ml in deionized water.

2.4 Measurement of the Mitochondrial Transmembrane Potential

2.4.1 Solutions and Reagent Stocks

1. 0.1 M phosphate buffer pH 7.4 (*see* **Note 4**): Directly before use 4.05 ml of Na_2HPO_4 solution (1.42 g of Na_2HPO_4 dissolved in 50 ml of deionized water) should be mixed with 0.95 ml of NaH_2PO_4 solution (1.39 g NaH_2PO_4 dissolved in 50 ml of deionized water) and supplemented with 5 ml of deionized water.

2. Measurement medium: 75 mM sucrose, 225 mM mannitol, 1 mM $MgCl_2$, 1 mM phosphate buffer, 25 mM Tris–HCl, pH 7.4. Store at 4 °C (*see* **Note 2**).

3. Safranin O stock: 5 mM Safranin O aqueous solution.

4. Glutamate stock: 0.5 M glutamate aqueous solution, pH 7.4 (adjusted with KOH).

5. Malate stock: 0.5 M malate aqueous solution, pH 7.4 (adjusted with KOH).

6. Rotenone stock: 1 mM rotenone ethanol solution.

7. Succinate stock: 0.5 M succinate aqueous solution, pH 7.4 (adjusted with KOH).

8. CCCP stock: 1 mM CCCP ethanol solution.

2.4.2 Equipment and Accessories

1. Spectrofluorimeter.

2. Acryl cuvettes $10 \times 10 \times 48$ mm for spectrofluorimeter.

2.5 Measurement of the Oxygen Consumption

2.5.1 Solutions and Reagent Stocks

1. 0.1 M phosphate buffer pH 7.4 (*see* **Note 4**): Directly before use 4.05 ml of Na_2HPO_4 solution (1.42 g of Na_2HPO_4 dissolved in 50 ml of deionized water) should be mixed with 0.95 ml of NaH_2PO_4 solution (1.39 g NaH_2PO_4 dissolved in 50 ml of deionized water) and supplemented with 5 ml of deionized water.

2. Measurement medium: 75 mM sucrose, 225 mM mannitol, 1 mM $MgCl_2$, 1 mM phosphate buffer, 25 mM Tris–HCl, pH 7.4. Store at 4 °C (*see* **Note 2**).

3. Glutamate stock: 0.5 M glutamate aqueous solution, pH 7.4 (adjusted with KOH).

4. Malate stock: 0.5 M malate aqueous solution, pH 7.4 (adjusted with KOH).

5. ADP stock: 100 mM ADP aqueous solution.

6. Oligomycin stock: 1 mM oligomycin ethanol solution.

2.5.2 Equipment

1. Clark-type oxygen electrode (e.g., YSI, Yellow Springs, OH, USA or Oroboros Oxygraph-2 k, Oxygraph®, Bioenergetics and Biomedical Instruments, Innsbruck, Austria) equipped with a unit calculating the rate of oxygen consumption (first derivative of the oxygen concentration trace).

---

3 Methods

To isolate an amount of mitochondria sufficient for functional studies 30–60 of confluent 10 cm dishes (depending on the cell type) have to be used. Less material is needed to isolate crude mitochondrial fraction for proteomic studies.

Measurements of the mitochondrial potential and oxygen consumption are often used to verify the quality of mitochondrial preparations.

3.1 Isolation of Crude Mitochondrial Fraction from Cell Lines

1. *Cells growing in monolayer*: Remove the medium, wash the cells twice with PBS (without Ca^{2+} and Mg^{2+}) and trypsinize them for 3–5 min in the incubator (37 °C) in order to detach cells. Add DMEM to stop trypsinization and centrifuge the content for 3 min at $200 \times g$ at room temperature.

2. *Cells growing in suspension*: Centrifuge the content of the cultivation flasks for 3 min at $200 \times g$ at room temperature.

3. Discard the supernatant and resuspend the cells in PBS (with Ca^{2+} and Mg^{2+}).

4. Centrifuge the resuspended cells for 3 min at $200 \times g$ at room temperature.

5. Resuspend the cell pellet in 20 ml of the ice-cold homogenization medium (*see* **Note 3**).

6. Homogenize the cells in a motor-driven tightly fitting glass/Teflon Potter–Elvehjem homogenizer.

7. Check the integrity of homogenized cells with a light microscope.

8. Centrifuge the homogenate for 5 min at $1,000 \times g$ at 4 °C.

9. Discard the pellet (unbroken cells and nuclei) and centrifuge the supernatant again for 5 min at $1,000 \times g$ at 4 °C.

10. Discard the pellet (remaining unbroken cells and nuclei) and centrifuge the supernatant for 10 min at $8,000 \times g$ at 4 °C.

11. Discard the supernatant (containing the cytosolic fraction, plasma membrane, lysosomes, and microsomes).

12. Gently resuspend the mitochondrial pellet in approximately 10 ml of the mitochondria isolation medium using a loose Potter–Elvehjem homogenizer and centrifuge it again for 10 min at $8,000 \times g$ at 4 °C (*see* **Note 4**).

13. Discard the supernatant and gently resuspend the mitochondrial pellet in 5–10 ml (depending on the pellet volume) of the mitochondria isolation medium using a loose Potter–Elvehjem homogenizer. Centrifuge it again for 10 min at $10,000 \times g$ at 4 °C.

14. Gently resuspend the crude mitochondrial pellet in ~2 ml of the mitochondria isolation medium using a loose Potter–Elvehjem homogenizer and store on ice.

15. The material can now be used for further functional experiments, e.g., measurement of oxygen consumption, measurement of mitochondrial membrane potential, production of reactive oxygen species as well as calcium uptake assay. In the case of proteomic analysis it could be necessary to perform additional purification of the crude fraction in order to obtain a pure mitochondrial fraction as described in [5].

3.2 Measurement of Protein Concentration

1. Protein assay is based on the Bradford method [9]. Fill a 3 ml spectrophotometer cuvette with 2.4 ml water. Then, depending on the density of the mitochondrial fraction, add a small sample (usually 10–50 μl) of the mitochondrial suspension (the resulting absorbance should optimally be 0.1–0.6). Then add 600 μl of room temperature Bio-Rad Protein Assay and shake the sample. Measure the absorbance at 595 nm. Repeat the same procedure with BSA standard solution using: 1.0, 2.0, 5.0, 10, and 20 μg protein per cuvette.

3.3 Measurement of the Mitochondrial Membrane Potential with the Use of Safranin O

1. Adjust the fluorometer: excitation wavelength, 495 nm; emission wavelength, 586 nm; slits (excitation and emission) ~3.

2. Fill the fluorometer cuvette with 3 ml of the measurement medium (*see* **Note 5**).

3. Add 3 μl of 5 mM Safranin O, 30 μl of 0.5 M glutamate, and 30 μl of 0.5 M malate stock solutions.

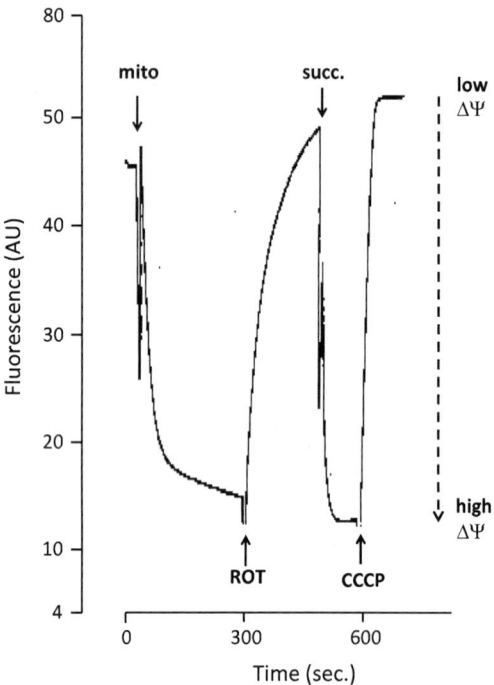

Fig. 1 Evaluation of quality of the crude mitochondrial fraction isolated from HeLa cells by measuring mitochondrial membrane potential (ΔΨ) with the use of Safranin O. Effect of rotenone and CCCP. Addition of rotenone (in the presence of glutamate and malate as substrates) leads to the collapse of ΔΨ due to the inhibition of complex I of the mitochondrial respiratory chain. Subsequent addition of succinate (complex II substrate) restores ΔΨ. Addition of CCCP (a protonophore) leads to the collapse of ΔΨ (uncoupling effect)

4. Start measuring fluorescence.

5. Add the mitochondrial suspension corresponding to about 1 mg protein and observe changes in the fluorescence. When it reaches a plateau continue with the following consecutive additions: 5 μl of 1 mM rotenone, next 30 μl of 0.5 M succinate and at the end 5 μl of 1 mM CCCP (to uncouple mitochondria) stock solutions.

An example of the results is shown in Fig. 1.

3.4 Measurement of Oxygen Consumption (Respiration)

1. Fill the chamber with the measurement medium (*see* **Note 5**).

2. Add 10 μl of 0.5 M glutamate and 10 μl of 0.5 M malate stock solutions.

3. Start running the oxygen concentration trace.

4. Add mitochondrial suspension (approximately 1 mg protein).

5. Wait for a stable signal (usually it takes about 1–3 min). Then continue measurement with the following consecutive additions: 5 μl of 100 mM ADP and later 2 μl of 1 mM oligomycin stock solutions.

An example of the results is shown in Fig. 2.

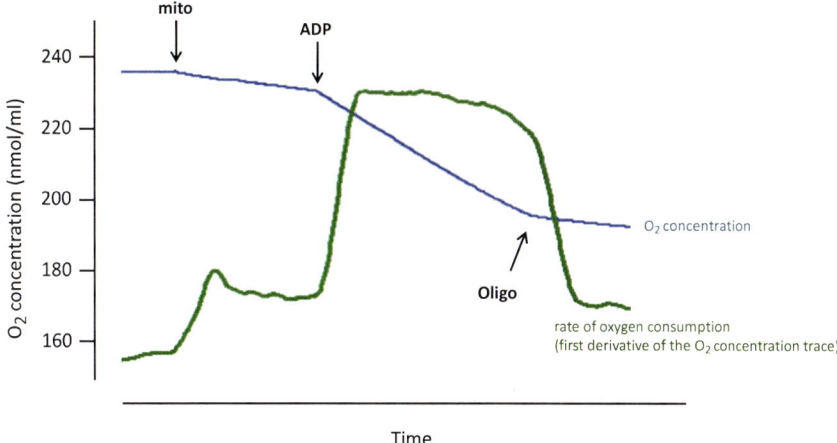

Fig. 2 Evaluation of quality of the crude mitochondrial fraction isolated from HeLa cells by measuring oxygen consumption. Effect of ADP and oligomycin. Mitochondrial respiration at low ADP level (state 4) is slow but increases dramatically after addition of exogenous ADP (state 3). Addition of oligomycin (oligo; inhibitor of ATP synthase) results in a decrease in oxygen consumption due to the inhibition of ATP production. Respiratory control ratio (RCR) is the quotient of state 3 to state 4. High RCR value indicates better integrity of the inner mitochondrial membrane and tighter coupling between the electron transport and ATP synthesis

4   Notes

1. To isolate intact mitochondria, it is necessary to use low-calcium sucrose (e.g., Merck, cat. no. 100892.9050).

2. The medium can be prepared in advance and stored at 4 °C for up to 2 weeks.

3. Homogenization and the following steps must be performed at 4 °C to minimize the activity of proteases and phospholipases. All solutions should be at 4 °C and all equipment pre-cooled.

4. Extreme care should be taken to avoid contamination with the ice and tap water.

5. For the assay the measurement media should have room temperature.

Acknowledgments

This work was supported by the Statutory Funding from Nencki Institute of Experimental Biology, Polish Ministry of Science and Higher Education grant W100/HFSC/2011.

References

1. Hogeboom GH, Schneider WC, Pallade GE (1948) Cytochemical studies of mammalian tissues. I. Isolation of intact mitochondria from rat liver; some biochemical properties of mitochondria and submicroscopic particulate material. J Biol Chem 172:619–635

2. Nessi P, Billesbolle S, Fornerod M, Maillard M, Frei J (1977) Leucocyte energy metabolism. VII. Respiratory chain enzymes, oxygen consumption and oxidative phosphorylation of mitochondria isolated from leucocytes. Enzyme 22:183–195

3. Moreadith RW, Fiskum G (1984) Isolation of mitochondria from ascites tumor cells permeabilized with digitonin. Anal Biochem 137: 360–367

4. Storrie B, Madden EA (1990) Isolation of subcellular organelles. Methods Enzymol 182: 203–225

5. Wieckowski MR, Giorgi C, Lebiedzinska M, Duszynski J, Pinton P (2009) Isolation of mitochondria-associated membranes and mitochondria from animal tissues and cells. Nat Protoc 4:1582–1590

6. Hornig-Do HT, Günther G, Bust M, Lehnartz P, Bosio A, Wiesner RJ (2009) Isolation of functional pure mitochondria by superparamagnetic microbeads. Anal Biochem 389:1–5

7. Bogucka K, Wroniszewska A, Bednarek M, Duszyński J, Wojtczak L (1990) Energetics of Ehrlich ascites mitochondria: membrane potential of isolated mitochondria and mitochondria within digitonin-permeabilized cells. Biochim Biophys Acta 1015:503–509

8. Pallotti F, Lenaz G (2007) Isolation and subfractionation of mitochondria from animal cells and tissue culture lines. Methods Cell Biol 80: 3–44

9. Bradford MM (1976) A rapid and sensitive method for the quantitation of microgram quantities of protein utilizing the principle of protein-dye binding. Anal Biochem 72:248–254

Chapter 2

Isolation of Mitochondria from Liver and Extraction of Total RNA and Protein: Analyses of MicroRNA and Protein Expressions

Pedro M. Borralho, Clifford J. Steer, and Cecília M.P. Rodrigues

Abstract

Several studies have indicated the presence of microRNAs (miRNAs) within mitochondria, although the origin, as well as the biological function of these mitochondrially located miRNAs is largely unknown. The identification and significance of this subcellular localization is gaining increasing relevance to the pathogenesis of certain disease states. Here we describe the isolation of highly purified mitochondria from rat liver by differential centrifugation, followed by RNAse A treatment to eliminate contaminating RNA. The coupled extraction of total RNA and protein is a more efficient design for allowing the downstream evaluation of miRNA and protein expression in mitochondria.

Key words Differential centrifugation, miRNA, Mitochondria, Protein, TRIzol

1 Introduction

Mitochondria are central organelles in the regulation of cellular homeostasis, playing a pivotal role in energy metabolism and cell viability, with their dysfunction or dysregulation being associated with multiple diseases. Human mitochondria harbor a compact circular genome of 16,569 bp in length, encoding 13 protein subunits of the electron transport chain, and a number of noncoding RNAs. Replication and transcription of mitochondrial DNA (mtDNA) are initiated from the small noncoding region known as the D loop, and regulated by proteins encoded in the nuclear genome, which are posttranslationally imported into the mitochondria [1]. In addition, transcription and translation of mtDNA, and mitochondrial transcript processing are regulated by several types of noncoding RNAs, either encoded in the mitochondrial genome, or in the nuclear genome following translocation to the mitochondria [2]. Interestingly, mitochondrial RNAs are transcribed from both strands as long polycistronic precursor transcripts, which undergo

Carlos M. Palmeira and Anabela P. Rolo (eds.), *Mitochondrial Regulation*, Methods in Molecular Biology,
vol. 1241, DOI 10.1007/978-1-4939-1875-1_2, © Springer Science+Business Media New York 2015

processing and final release of both noncoding and coding RNAs, including tRNAs, rRNAs, and mRNAs.

In the last two decades, microRNAs (miRNAs) have emerged as a class of noncoding RNAs processed from endogenous transcripts, functioning as master regulators and fine-tuners of the genome, via posttranscriptional mechanisms [3]. miRNA gene-silencing occurs via miRNA-mRNA binding, typically with incomplete albeit high sequence complementarity between the 3′-untranslated region target sites of mRNA transcripts and the miRNA seed sequence located at position 2–7 of its 5′-end [4]. This allows miRNAs to regulate the expression of multiple target genes and signaling pathways involved in the regulation of key cellular processes including cell growth, proliferation, differentiation, and apoptosis [5] as well as mitochondrial function [2]. Consequently, deregulation of a given miRNA can lead to malfunction of pivotal cellular mechanisms, contributing to disease onset and/or progression. Therefore, it is not surprising that miRNA modulation is increasingly demonstrated as a relevant therapeutic strategy in human disease [3, 6–9].

Since their discovery, miRNAs have been identified in multiple biological systems. Currently, almost 2,600 mature human miRNAs have been identified (mirbase.org), and predicted to regulate at least 60 % of protein-coding genes. Recent studies, including one by our group have demonstrated the presence of mature miRNAs not only in the cytoplasm, where target gene silencing of the mRNA transcript occurs, but also in other cellular compartments, including the nucleus [10], nucleolus [11, 12] and mitochondria [13–16]. In addition to mature miRNAs, precursor miRNAs have also been found in mitochondria [15]. However, it is not known whether mitochondrial localized miRNAs are encoded in the mitochondrial genome or in the nuclear genome and later imported into the mitochondria. Nevertheless, and regardless of their provenance from the nuclear or mitochondrial genome, the biological and pathophysiological implications of this unexpected finding are incompletely explored. Based on recent studies, it is conceivable that mitochondria-localized miRNAs may putatively function as posttranscriptional regulators, or fine-tuners, of the mitochondrial genome. This is illustrated by the mitochondrial translocation of the nuclear encoded miR-181c, which was shown to regulate the expression of the mitochondria-encoded cytochrome c oxidase subunit 1 (mt-cox1) in rat cardiac myocytes [17].

The first evidence that RNAi components may localize to the mitochondria was provided almost a decade ago, by the demonstration that human Ago2 interacts with mitochondrial tRNA[met] [18]. Interestingly, miRNAs were later identified in mouse liver mitochondria by small RNA sequencing [19]. Aside from this unexpected finding, it was proposed that the presence of miRNAs in mouse liver mitochondria may arise from cytosolic contamination of the isolated mitochondria. Subsequently, we have overcome this technical limitation by treating isolated and purified rat liver mitochondria with RNAse A, to ensure that mitochondria are devoid of contaminant RNAs from the cytosol or other cellular organelles and compartments. Coupling this extremely relevant purification step to the detection of mature miRNAs by microarray followed by northern blot confirmation, we have clearly demonstrated that mature miRNAs are indeed present in highly purified mitochondria [13]. Since then, additional reports using RNAse treatment and microarray methodologies, confirmed the detection of miRNAs in purified mitochondria isolated from mouse liver [14], and from human cervical adenocarcinoma (HeLa) cells [16].

The identification of miRNAs in mitochondria and the exploration of their biological role in this subcellular organelle have been gaining increasing relevance. Mitochondria may be easily isolated and purified from multiple organs or cultured cells. The methodologies described in this chapter allow the straightforward isolation of highly purified mitochondria, and the coupled extraction of total RNA and protein, allowing downstream evaluation and comparison of mitochondrial miRNA and protein expression, in the same sample.

2   Materials

Prepare all solutions using ultrapure water (prepared by purifying deionized water to attain a sensitivity of 18 MΩ cm at 25 °C) and analytical grade reagents. Prepare and store all reagents at 4 °C (unless otherwise indicated). Diligently follow all waste disposal regulations when disposing waste materials.

2.1 Mitochondrial Isolation: Materials and Reagents

1. *Speed controlled mechanical drill*: Tri-R model K41 skill drill (Tri-R Instruments).

2. *Tissue grinder*: Glass mortar with radially serrated PTFE pestle (Fischer Scientific).

3. *Ultra-spin buffer*: 0.25 M Sucrose, 1 mM Ethylene glycol-bis (2-aminoethylether)- N,N,N',N'-tetraacetic acid (EGTA). Add 150 ml of water to a graduated cylinder. Weigh 21.39 g sucrose and 95.1 mg EGTA, and transfer to the graduated cylinder. Mix and adjust pH to 7.4 with KOH. Complete volume with water to 250 ml (*see* **Note 1**).

4. *PERCOLL* (Colloidal PVP coated silica for cell separation) (#P1644, Sigma-Aldrich).

5. *Homogenate buffer*: 70 mM sucrose, 220 mM mannitol, 1 mM EGTA, 10 mM 4-(2-Hydroxyethyl)piperazine-1-ethanesulfonic acid, N-(2-Hydroxyethyl)piperazine- N'-(2-ethanesulfonic acid) (HEPES), pH 7.4. Add 700 ml of water to a glass beaker.

Weigh 23.6 g sucrose, 40.8 g mannitol, 380.4 mg EGTA, and 2.38 g HEPES and transfer to the cylinder. Mix and adjust pH to 7.4 with KOH. Adjust volume to 1 l.

6. *Mitochondria wash buffer 1*: 0.1 M KCl, 5 mM 3-[N-Morpholino]propanesulfonic acid (MOPS), 1 mM EGTA. Add 800 ml of water to a glass beaker. Weigh 7.46 g KCl, 1.05 g MOPS, and 0.380 mg EGTA, and transfer to the glass beaker. Mix and adjust pH to 7.4 with KOH. Adjust volume to 1 l.

7. *Mitochondria wash buffer 2*: 0.1 M KCl, 5 mM MOPS. Add 800 ml of water to a glass beaker. Weigh 7.46 g KCl, and 1.05 g MOPS, and transfer to the glass beaker. Mix and adjust pH to 7.4 with KOH. Adjust volume to 1 l.

8. *Mitochondria suspension buffer*: 125 mM Sucrose, 50 mM KCl, 5 mM HEPES, 2 mM KH_2PO_4. Add 70 ml of water to a glass beaker. Weigh 4.28 g Sucrose, 372.8 mg KCl, 119.15 mg HEPES, and 27 mg KH_2PO_4 and transfer to the glass beaker. Mix and adjust volume to 100 ml.

9. *EDTA-free Complete®-Mini protease inhibitor cocktail* (#04693159001, Roche Applied Science).

10. *Chelex® 100 Resin* (#143-2832, Bio-Rad Laboratories, CA, USA).

11. *RNAse A*, 20 mg/ml in 50 mM Tris-HCl (pH 8.0), 10 mM EDTA (#12091-039, Invitrogen).

2.2 Total RNA Isolation: Materials and Reagents

1. *NanoDrop* (Thermo Scientific).

2. *TRIzol®* (#15596-026, Life Technologies).

3. *Chloroform*.

4. *Isopropanol*.

5. *UltraPure™ DNase/RNase-Free Distilled Water* (#10977-023, Invitrogen).

6. *75 % (vol/vol) Ethanol*. Add 15 ml of Ethanol and 5 ml of UltraPure™ DNase/RNase-Free Distilled Water in a sterile tube, and mix well. Store at room temperature.

2.3 Total Protein Isolation: Materials and Reagents

1. *Compact Ultrasonic Device*: model UP100H—100 W, ultrasonic frequency 30 kHz (Hielscher Ultrasonics GmbH, Teltow, Germany).

2. *Tris–Hcl 1 M, pH 8.0* (#T2694, Sigma-Aldrich).

3. *Protein wash buffer*: 0.3 M guanidine hydrochloride in 95 % ethanol. Add 142.5 ml of Ethanol and 7.5 ml of water to a graduated cylinder to prepare 150 ml of 95 % ethanol. Add 80 ml of 95 % ethanol to a beaker. Weigh 2.87 g of guanidine hydrochloride and transfer to the beaker. Bring the volume to

100 ml using 95 % ethanol, and mix well. Store at room temperature.

4. *1 % (wt/vol) Sodium dodecyl sulfate (SDS)*: Add 80 ml of water to a graduated cylinder. Weigh 1 g of SDS and transfer to the cylinder. Mix well and bring the volume to 100 ml. Store at room temperature.

5. *8 M Urea in Tris–HCl 1 M, pH 8.0*: Add 80 ml of water to a graduated cylinder. Weigh 48.05 g of Urea and transfer to the cylinder. Mix well and bring the volume to 100 ml. Store at room temperature.

6. *Protein resuspension buffer*: 1:1 (vol/vol) of 1 % SDS and 8 M urea in Tris–HCl 1 M, pH 8.0. Add 25 ml of 1 % SDS solution and 25 ml of 8 M urea in Tris–HCl 1 M, pH 8.0 to a glass flask, as 50 ml of Protein Resuspension Buffer. Mix well and store at room temperature.

3 Methods

3.1 Mitochondria Isolation

The flowchart in Fig. 1 depicts the protocol followed after preparing the buffers and obtaining liver fragments.

1. Prepare two self-generating gradients (ultra-spin buffer:Percoll (75:25); vol/vol) by adding 15 ml of ultra-spin buffer and 5 ml of PERCOLL to each ultracentrifuge tube. Mix well and centrifuge the tubes at $43,000 \times g$ for 30 min, at 4 °C. Carefully store the gradients at 4 °C until use (*see* **Note 2**).

2. Prepare homogenate buffer containing EDTA-free Complete®-Mini protease inhibitor cocktail, by adding 12 tablets to 120 ml of homogenate buffer. Mix to dissolution and place on ice (*see* **Note 3**).

3. Prepare 1 % (wt/vol) Chelex-100-treated mitochondria wash buffer 2. Add 1.5 g of Chelex to 150 ml of mitochondria wash buffer 2. Mix well by vigorous shaking by hand and incubate for 10 min, at room temperature (*see* **Note 4**). Place on ice until use.

4. Prepare 1 % (wt/vol) Chelex-100-treated mitochondria suspension buffer. Add 100 mg of Chelex to 10 ml of suspension buffer. Mix well by vigorous shaking by hand, and incubate for 10 min at room temperature (*see* **Note 4**). Place on ice until use.

5. Fill three glass petri dishes with 50 ml of ice-cold homogenate buffer, and place them on ice, together with an additional empty glass plate.

6. Sacrifice an adult male 175–200 g Sprague-Dawley or Wistar rat by exsanguination under CO_2 anesthesia.

Homogenize the sample as a 10% (wt/vol) homogenate.
in homogenate buffer with protease inhibitor.
Use 6-10 up & down strokes with speed controlled
mechanical drill and tissue grinder, at 800 rpm

↓ 600 g, 10 min, 4° C

Collect the supernatant (discard the pellet)

↓ 1,100 g, 10 min, 4° C

Collect the supernatant (discard the pellet)

↓ 7,700 g, 10 min, 4° C

Collect the pellet (discard the supernatant)

↓

Resuspend the pellet in homogenate buffer with protease inhibitor.
Carefully and slowly layer on top of self-generating gradient

↓ 43,000 g, 1 h, 4° C

Collect the lower yellowish-brown mitochondrial band.
Resuspend in mitochondria wash buffer 1

↓ 7,700 g, 10 min, 4° C

Wash purified mitochondria twice by
resuspending in chelex-100-treated mitochondria wash buffer 2

↓ 7,700 g, 10 min, 4° C

Collect the pellet (discard the supernatant)

↓

Resuspend in chelex-100-treated mitochondria suspension buffer.
Add 2 mg of RNAse A per ml.

↓ Incubate 1h, 37 ° C
12,000 g, 3 min, 4° C

Collect the pellet (discard the supernatant)

↓

Wash mitochondria pellet twice by resuspending in
chelex-100-treated mitochondria suspension buffer

↓ 12,000 g, 3 min, 4° C

Collect the mitochondrial pellet (discard supernatant)
and proceed to RNA extraction (section 3.2).
(alternatively you may store the mitochondria at -80° C)

Fig. 1 Flowchart of mitochondrial isolation. Includes all steps following the preparation of buffers and collecting the liver sample for processing

7. Remove the rat liver, and using tweezers, rinse the liver in the three petri dishes containing ice-cold homogenate buffer (prepared in **step 5**), and keep the rat liver on ice in the final rinse solution (*see* **Note 5**).

8. Using the ice-cold empty glass plate and a laboratory scale, weigh approximately 10 g of rat liver, and mince the sample into small fragments using scissors (*see* **Note 6**).

9. Homogenize the sample as a 10 % (wt/vol) homogenate. Homogenize the 10 g of minced liver in 100 ml ice-cold homogenate buffer containing EDTA-free Complete®-Mini protease inhibitor cocktail (prepared in **step 2**), using 6–10 complete up and down strokes with a speed controlled mechanical drill and tissue grinder, at 800 rpm. Place the homogenate on ice.

10. Centrifuge the homogenate at $600 \times g$ for 10 min, at 4 °C. Collect the supernatant and discard the pellet (*see* **Note 7**).

11. Centrifuge the supernatant at $1,100 \times g$ for 10 min, at 4 °C. Collect the supernatant and discard the pellet (*see* **Note 8**).

12. Centrifuge the supernatant at $7,700 \times g$ for 10 min, at 4 °C. Discard the supernatant and keep the pellet (*see* **Note 9**).

13. Resuspend the pellet in 4 ml of homogenate buffer containing EDTA-free Complete®-Mini protease inhibitor cocktail (prepared in **step 2**), to obtain a crude mitochondrial extract.

14. Carefully, and slowly, layer 2 ml of the crude mitochondrial extract to each of the 20 ml of self-generating gradients (prepared in **step 1**). To purify the mitochondria, centrifuge at $43,000 \times g$ for 1 h, at 4 °C.

15. Carefully collect the lower yellowish-brown mitochondrial band and resuspend each in 30 ml of mitochondrial wash buffer 1.

16. Centrifuge at $7,700 \times g$ for 10 min, at 4 °C. Discard the supernatant and keep the purified mitochondrial pellet.

17. Perform two sequential washes of the purified mitochondria, by resuspending the mitochondrial pellet in 30 ml of Chelex-100-treated mitochondrial wash buffer 2 (prepared in **step 3**) (*see* **Note 10**), followed by centrifugation at $7,700 \times g$ for 10 min, at 4 °C, discarding the supernatant and keeping the pellet.

18. Resuspend and pool the purified mitochondrial samples in 1–2 ml of Chelex-100-treated mitochondrial suspension buffer (prepared in **step 4**).

19. Perform RNAse treatment by adding 2 mg of RNAse A per ml of purified mitochondrial suspension, and incubate at 37 °C for 1 h (*see* **Note 11**).

20. Centrifuge the mitochondrial suspension at $12,000 \times g$ for 3 min, at 4 °C. Discard supernatant and maintain the purified mitochondrial pellet on ice.

21. Wash the mitochondrial pellet two times by resuspending in Chelex-100-treated mitochondrial suspension buffer (prepared in **step 4**) (*see* **Note 10**), and centrifuging at $12,000 \times g$ for 3 min, at 4 °C.

22. Discard the supernatant and place the purified mitochondria on ice. Proceed directly to total RNA extraction (go to Subheading 3.2) (*see* **Note 12**).

3.2 Total RNA Extraction from Isolated Mitochondria

The flowchart in Fig. 2 depicts the protocol followed.

1. Add TRIzol® to the purified mitochondria pellet, and pipette up and down to homogenize the sample (*see* **Note 13**). Incubate for 5 min at room temperature.

2. Add 0.2 ml of Chloroform per ml of TRIzol® reagent used for sample homogenization, and vigorously shake, by hand, for 15 s, followed by a 3-min incubation at room temperature.

3. Centrifuge at $12,000 \times g$ for 15 min, at 4 °C to allow phase separation, and collect the upper aqueous phase into a new tube (*see* **Note 14**). The interphase and lower phenol–chloroform phase may be stored at 4 °C, or at –80 °C for prolonged storage, and subsequent total protein extraction (Subheading 3.3) (*see* **Note 15**).

4. Precipitate the RNA by adding 0.5 ml of Isopropanol per ml of TRIzol® reagent used for sample homogenization to the aqueous phase containing the RNA. Shake by hand for 15 s and incubate at room temperature for 10 min (*see* **Note 16**).

5. Centrifuge at $12,000 \times g$ for 10 min, at 4 °C, and discard the supernatant.

6. Wash the RNA pellet by adding 1 ml of 75 % Ethanol per ml of TRIzol® reagent used for sample homogenization, and mix the tube by inversion ×10.

7. Centrifuge at $7,500 \times g$ for 5 min, at 4 °C, and discard the supernatant.

8. Air-dry the RNA pellet for 5 min (*see* **Note 17**).

9. Resuspend the RNA in UltraPure™ DNase/RNase-Free Distilled Water and quantitate the RNA using NanoDrop 1000 (Thermo Scientific) (*see* **Note 18**).

10. Proceed to downstream applications (*see* **Note 19**) or store the RNA at –80 ° C.

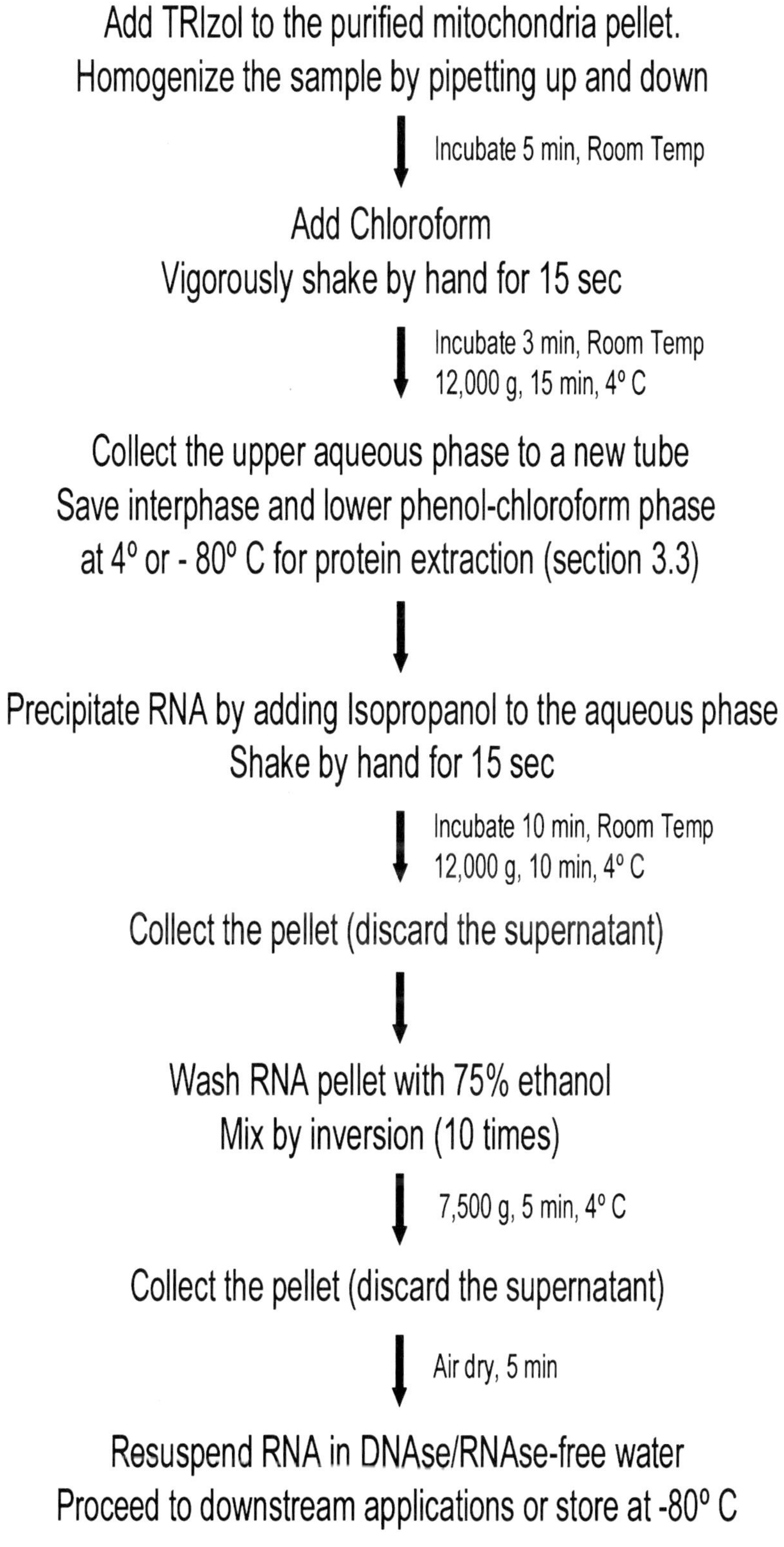

Add TRIzol to the purified mitochondria pellet.
Homogenize the sample by pipetting up and down

↓ Incubate 5 min, Room Temp

Add Chloroform
Vigorously shake by hand for 15 sec

↓ Incubate 3 min, Room Temp
12,000 g, 15 min, 4° C

Collect the upper aqueous phase to a new tube
Save interphase and lower phenol-chloroform phase
at 4° or - 80° C for protein extraction (section 3.3)

↓

Precipitate RNA by adding Isopropanol to the aqueous phase
Shake by hand for 15 sec

↓ Incubate 10 min, Room Temp
12,000 g, 10 min, 4° C

Collect the pellet (discard the supernatant)

↓

Wash RNA pellet with 75% ethanol
Mix by inversion (10 times)

↓ 7,500 g, 5 min, 4° C

Collect the pellet (discard the supernatant)

↓ Air dry, 5 min

Resuspend RNA in DNAse/RNAse-free water
Proceed to downstream applications or store at -80° C

Fig. 2 Flowchart of total RNA extraction from isolated mitochondria

3.3 Total Protein Extraction from TRIzol®-Treated Isolated Mitochondria

The flowchart in Fig. 3 depicts the protocol followed.

1. Start with the interphase and lower phenol–chloroform phase obtained after RNA isolation (**step 3** of Subheading 3.2), and stored either at 4 °C or at –80 °C. If necessary, thaw the samples at room temperature, and centrifuge at $12,000 \times g$ for 15 min, at 4 °C. Discard the remaining upper aqueous phase.

2. Add 0.3 ml of 100 % Ethanol per ml of TRIzol® reagent used for sample homogenization to precipitate DNA. Mix by inversion and incubate for 3 min at room temperature.

3. Centrifuge at $2,000 \times g$ for 5 min, at 4 °C. Transfer the phenol–ethanol supernatant to a new tube for protein isolation (*see* **Note 20**).

4. Add 1.5 ml of Isopropanol per ml of TRIzol® reagent used for sample homogenization. Mix by inversion and incubate for 10 min, at room temperature.

5. Centrifuge at $12,000 \times g$ for 10 min, at 4 °C to precipitate protein, and discard the supernatant.

6. Perform three sequential washes of the protein pellets by adding 2 ml of protein wash buffer per ml of TRIzol® reagent used for sample homogenization, vigorously shaking the tubes by hand, incubating at room temperature for 20 min, and centrifuging at $7,500 \times g$ for 5 min, at 4 °C (*see* **Note 21**).

7. After the third wash and spin, add 2 ml of 100 % ethanol per ml of TRIzol® reagent, incubate at room temperature for 20 min, and centrifuge at $7,500 \times g$ for 5 min at, 4 °C (*see* **Note 21**). Discard the supernatant and maintain the protein pellet on ice (*see* **Note 22**).

8. Solubilize the protein pellet by adding protein resuspension buffer, followed by five cycles of 15 s sonication and 30 s ice incubation, using a compact ultrasonic device with amplitude adjusted to 80 % and pulse to 90 %.

9. Centrifuge at $3,200 \times g$ for 10 min at 4 °C, to sediment insoluble material. Discard the pellet, and collect the supernatant containing the solubilized proteins to a fresh tube and store at –80 °C for downstream applications.

4 Notes

1. Sucrose solutions are easily contaminated, even when stored at 4 °C. Use the solution within 1 week. Ideally, prepare just before use.

2. The gradients should always be prepared prior to use, and carefully handled to avoid mixing and discontinuities that may alter the efficiency of the separations.

Fresh or frozen (-80º C) interphase and lower
phenol-chloroform phase from total RNA extraction from
isolated mitochondria (section 3.2)
(Thaw frozen samples)

↓ 12,000 g, 15 min, 4º C

Discard remaining upper aqueous phase
Precipitate DNA by adding 100% ethanol
Mix by inversion

↓ Incubate 3 min, Room Temp
2,000 g, 5 min, 4º C

Collect the phenol-ethanol supernatant (discard pellet)
Precipitate protein with isopropanol
Mix by inversion

↓ Incubate 10 min, Room Temp
12,000 g, 10 min, 4º C

Collect the protein pellet (discard the supernatant)

↓

Wash the protein pellet three times with protein wash buffer
Vigorously shake by hand

↓ Incubate 20 min, Room Temp
7,500 g, 5 min, 4º C

Wash protein pellet with 100% ethanol
Vigorously shake by hand

↓ Incubate 20 min, Room Temp
7,500 g, 5 min, 4º C

Solubilize the protein pellet in protein resuspension buffer
Sonicate the sample

↓ 3,200 g, 10 min, 4º C

Collect the clear supernatant containing the protein
and store at -80º C for downstream applications

Fig. 3 Flowchart of total protein extraction from TRIzol®-treated isolated mitochondria

3. EDTA-free Complete®-Mini protease inhibitor cocktail should be added before use, and the buffer kept on ice. Always prepare fresh, discarding remainder of the solution, to ensure maximal protease inhibition.

4. Treatment of buffers with chelating resin Chelex-100 should be performed before use. Discard the remainder of Chelex-100 treated buffers, to ensure optimal clearing of the buffers.

5. Rinsing in three consecutive changes of homogenate buffer eliminates blood and other contaminants, which may be located on the outer surface of the liver during animal sacrifice and organ collection. After the final rinse of the liver in ice-cold homogenate buffer, the buffer should be clear and devoid of blood and macroscopic debris. If necessary, perform additional rinses in similar fashion.

6. Additional liver fragments may be stored for subsequent analysis (e.g., protein and/or RNA extraction from whole liver), by snap-freezing the samples in liquid nitrogen, followed by storage at –80 ° C.

7. This step clears the sample from insoluble tissue, incompletely lysed cells and other debris.

8. This step further separates the sample from insoluble tissue, incompletely lysed cells and other debris.

9. This step precipitates mitochondria and other cellular organelles.

10. The Chelex-100 resin precipitates by gravity. Following the 10 min incubation at room temperature, the Chelex-100 resin deposits at the bottom of the solution. Pipette the buffer without disturbing the Chelex-100 resin deposit.

11. RNAse treatment of isolated mitochondria eliminates RNAs in solution and those located on the outer surface of the mitochondria, ensuring the absence of non-mitochondrial contaminating RNAs in the sample. Optionally, before and/or after RNAse A treatment, you may take a sample of purified mitochondria to evaluate mitochondria morphology and purity by transmission electron microscopy and western blot, as previously described [13, 20].

12. You may optionally snap-freeze the purified mitochondria in liquid nitrogen, followed by storage at –80 °C.

13. Isolation of total RNA using TRIzol® allows the isolation of microRNAs, suitable for multiple downstream applications [6–9, 21–23]. Alternatively, samples may be homogenized in TRIzol® reagent using a motor-driven Bio-vortexer (No1083; Biospec Products, Bartlesfield, OK) and disposable RNAse/DNAse free sterile pestles (Thermo Fisher Scientific, Inc., Chicago, IL) [8].

14. The sample mixture separates into an upper colorless aqueous phase, an interphase, and a lower red phenol–chloroform phase. The RNA containing the microRNAs is exclusively in the upper colorless aqueous phase.

15. Phenol–chloroform phases may be stored at 4 °C overnight or at –80 °C for long-term storage, of at least 2 years [24].

16. Alternatively, you may incubate the samples at –20 °C overnight to increase RNA precipitation, or as a stop point in the protocol.

17. Do not allow the RNA pellet to completely dry as this reduces its solubility.

18. Alternatively, you can quantitate the RNA by diluting the sample in UltraPure™ DNase/RNase-Free Distilled Water and determining its absorbance at 260 and 280 nM. RNA concentration may be determined using the formula: $A260 \times \text{dilution} \times 40 = \mu g$ of RNA per ml.

19. Downstream applications may involve, among others, evaluation of microRNA expression by microarray analysis [6, 13, 22, 23], northern blotting [21], or Taqman real-time RT-PCR [7–9].

20. This step precipitates DNA, with the proteins remaining in the phenol–ethanol supernatant.

21. It is not necessary to resuspend the protein pellet. Vigorously shake by hand and incubate at room temperature. Following centrifugation, completely remove each buffer solution to maximize wash efficacy.

22. Alternatively, the protein pellets may be snap-frozen in liquid nitrogen and stored at –80 °C until resuspension.

Acknowledgements

This work was supported by grant PTDC/SAU-ORG/119842/2010 from Fundação para a Ciência e a Tecnologia (FCT), and by grant from Sociedade Portuguesa de Gastrenterologia, Portugal.

References

1. Mercer TR, Neph S, Dinger ME, Crawford J, Smith MA, Shearwood AM, Haugen E, Bracken CP, Rackham O, Stamatoyannopoulos JA, Fili-povska A, Mattick JS (2011) The human mitochondrial transcriptome. Cell 146:645–658

2. Tomasetti M, Neuzil J, Dong L (2014) MicroRNAs as regulators of mitochondrial function: Role in cancer suppression. Biochim Biophys Acta 1840:1441–1453

3. Pereira DM, Rodrigues PM, Borralho PM, Rodrigues CM (2013) Delivering the promise of miRNA cancer therapeutics. Drug Discov Today 18:282–289

4. Pasquinelli AE (2012) MicroRNAs and their targets: recognition, regulation and an emerging

reciprocal relationship. Nat Rev Genet 13: 271–282

5. Huntzinger E, Izaurralde E (2011) Gene silencing by microRNAs: contributions of translational repression and mRNA decay. Nat Rev Genet 12:99–110

6. Oberg AL, French AJ, Sarver AL, Subramanian S, Morlan BW, Riska SM, Borralho PM, Cunningham JM, Boardman LA, Wang L, Smyrk TC, Asmann Y, Steer CJ, Thibodeau SN (2011) MiRNA expression in colon polyps provides evidence for a multihit model of colon cancer. PLoS One 6:e20465

7. Borralho PM, Kren BT, Castro RE, da Silva IB, Steer CJ, Rodrigues CM (2009) MicroRNA-143 reduces viability and increases sensitivity to 5-fluorouracil in HCT116 human colorectal cancer cells. FEBS J 276:6689–6700

8. Borralho PM, Simoes AE, Gomes SE, Lima RT, Carvalho T, Ferreira DM, Vasconcelos MH, Castro RE, Rodrigues CM (2011) MiR-143 overexpression impairs growth of human colon carcinoma xenografts in mice with induction of apoptosis and inhibition of proliferation. PLoS One 6:e23787

9. Castro RE, Ferreira DM, Afonso MB, Borralho PM, Machado MV, Cortez-Pinto H, Rodrigues CM (2013) MiR-34a/SIRT1/p53 is suppressed by ursodeoxycholic acid in the rat liver and activated by disease severity in human non-alcoholic fatty liver disease. J Hepatol 58:119–125

10. Park CW, Zeng Y, Zhang X, Subramanian S, Steer CJ (2010) Mature microRNAs identified in highly purified nuclei from HCT116 colon cancer cells. RNA Biol 7:606–614

11. Politz JC, Hogan EM, Pederson T (2009) MicroRNAs with a nucleolar location. RNA 15:1705–1715

12. Li ZF, Liang YM, Lau PN, Shen W, Wang DK, Cheung WT, Xue CJ, Poon LM, Lam YW (2013) Dynamic localisation of mature microRNAs in human nucleoli is influenced by exogenous genetic materials. PLoS One 8:e70869

13. Kren BT, Wong PY, Sarver A, Zhang X, Zeng Y, Steer CJ (2009) MicroRNAs identified in highly purified liver-derived mitochondria may play a role in apoptosis. RNA Biol 6:65–72

14. Bian Z, Li LM, Tang R, Hou DX, Chen X, Zhang CY, Zen K (2010) Identification of mouse liver mitochondria-associated miRNAs and their potential biological functions. Cell Res 20:1076–1078

15. Barrey E, Saint-Auret G, Bonnamy B, Damas D, Boyer O, Gidrol X (2011) Pre-microRNA and mature microRNA in human mitochondria. PLoS One 6:e20220

16. Bandiera S, Ruberg S, Girard M, Cagnard N, Hanein S, Chretien D, Munnich A, Lyonnet S, Henrion-Caude A (2011) Nuclear outsourcing of RNA interference components to human mitochondria. PLoS One 6:e20746

17. Das S, Ferlito M, Kent OA, Fox-Talbot K, Wang R, Liu D, Raghavachari N, Yang Y, Wheelan SJ, Murphy E, Steenbergen C (2012) Nuclear miRNA regulates the mitochondrial genome in the heart. Circ Res 110: 1596–1603

18. Maniataki E, Mourelatos Z (2005) Human mitochondrial tRNAMet is exported to the cytoplasm and associates with the Argonaute 2 protein. RNA 11:849–852

19. Lung B, Zemann A, Madej MJ, Schuelke M, Techritz S, Ruf S, Bock R, Huttenhofer A (2006) Identification of small non-coding RNAs from mitochondria and chloroplasts. Nucleic Acids Res 34:3842–3852

20. Rodrigues CM, Ma X, Linehan-Stieers C, Fan G, Kren BT, Steer CJ (1999) Ursodeoxycholic acid prevents cytochrome c release in apoptosis by inhibiting mitochondrial membrane depolarization and channel formation. Cell Death Differ 6:842–854

21. Castro RE, Ferreira DM, Zhang X, Borralho PM, Sarver AL, Zeng Y, Steer CJ, Kren BT, Rodrigues CM (2010) Identification of microRNAs during rat liver regeneration after partial hepatectomy and modulation by ursodeoxycholic acid. Am J Physiol Gastrointest Liver Physiol 299:G887–G897

22. Cunningham JM, Oberg AL, Borralho PM, Kren BT, French AJ, Wang L, Bot BM, Morlan BW, Silverstein KA, Staggs R, Zeng Y, Lamblin AF, Hilker CA, Fan JB, Steer CJ, Thibodeau SN (2009) Evaluation of a new high-dimensional miRNA profiling platform. BMC Med Genomics 2:57

23. Sarver AL, French AJ, Borralho PM, Thayanithy V, Oberg AL, Silverstein KA, Morlan BW, Riska SM, Boardman LA, Cunningham JM, Subramanian S, Wang L, Smyrk TC, Rodrigues CM, Thibodeau SN, Steer CJ (2009) Human colon cancer profiles show differential microRNA expression depending on mismatch repair status and are characteristic of undifferentiated proliferative states. BMC Cancer 9:401

24. Simoes AE, Pereira DM, Amaral JD, Nunes AF, Gomes SE, Rodrigues PM, Lo AC, D'Hooge R, Steer CJ, Thibodeau SN, Borralho PM, Rodrigues CM (2013) Efficient recovery of proteins from multiple source samples after TRIzol((R)) or TRIzol((R))LS RNA extraction and long-term storage. BMC Genomics 14:181

Chapter 3

PCR Based Determination of Mitochondrial DNA Copy Number in Multiple Species

John P. Rooney, Ian T. Ryde, Laurie H. Sanders, Evan. H. Howlett, Meryl D. Colton, Kaylyn E. Germ, Greg D. Mayer, J. Timothy Greenamyre, and Joel N. Meyer

Abstract

Mitochondrial DNA (mtDNA) copy number is a critical component of overall mitochondrial health. In this chapter, we describe methods for isolation of both mtDNA and nuclear DNA (nucDNA) and measurement of their respective copy numbers using quantitative PCR. Methods differ depending on the species and cell type of the starting material and availability of specific PCR reagents.

Key words Mitochondrial DNA, mtDNA, mtDNA depletion, Copy number, QPCR, Mitochondrial toxicity, Mitochondrial disease

1 Introduction

In humans, the mitochondrial genome (mtDNA) encodes 13 proteins, all of which are components of the electron transport chain (ETC), and are essential for oxidative phosphorylation (OXPHOS) [1]. On average, each cell contains between 10^3 and 10^4 copies of the mitochondrial genome, though this number varies between cell type and developmental stage. mtDNA replication is carried out independent of the cell cycle by the nuclear DNA (nucDNA) encoded polymerase γ, the only DNA polymerase found in the mitochondria [2]. mtDNA replication is regulated, at least in part, by Mitochondrial transcription factor A (Tfam), a component of the mtDNA-protein packaging complex called the nucleoid [3]. Mutations in nine genes involved in mtDNA replication and nucleotide metabolism cause mitochondrial DNA depletion syndrome (MDS) in humans [4–6]. There are multiple diseases associated with MDS including both Alper's syndrome and progressive external ophthalmoplegia (PEO), as well as other recessive myopathies [4]. mtDNA depletion is also implicated in more common diseases

Carlos M. Palmeira and Anabela P. Rolo (eds.), *Mitochondrial Regulation*, Methods in Molecular Biology, vol. 1241, DOI 10.1007/978-1-4939-1875-1_3, © Springer Science+Business Media New York 2015

including type 2 diabetes [7], many cancers [8], and neurodegenerative disorders such as Alzheimer's and Parkinson's Disease [9], though direct causal links have not yet been established. Pharmaceuticals can also block mtDNA replication and result in mtDNA depletion, as in the case of the Nucleoside Reverse Transcriptase Inhibitors (NRTIs) used to treat human immune-deficiency virus (HIV). These drugs are nucleoside analogs that block the progression of pol-γ and can cause mtDNA depletion mediated toxicity [10]. Furthermore, recent work suggests that environmental exposures can alter mtDNA copy number. mtDNA copy number is increased by in utero and neonatal exposure to second-hand cigarette smoke in mice [11] and chronic exposure to polycyclic aromatic hydrocarbons in humans [12]. Conversely, exposure to particulate matter during pregnancy reduces mtDNA copy number in placental tissue [13].

This chapter provides protocols for the simultaneous isolation of mitochondrial and nuclear DNA and measurement of both genome copy numbers from a variety of species. We present three protocols for mtDNA copy number determination. They differ based on availability and optimization of reagents. In *C. elegans*, actual mtDNA copy number per animal can be measured via real time PCR using a plasmid based mtDNA copy number standard curve [14]. With human samples, real time PCR can also be used; however, without a standard curve, such that the measurement of mtDNA copy number is relative to nuclear DNA copy number. A quantitative, non-real time PCR may also be used to measure relative mtDNA copy number in *Drosophila melanogaster*, *Fundulus heteroclitus* (Atlantic Killifish), *Fundulus grandis* (Gulf Killifish), *Danio rerio* (Zebrafish), *Oryzias latipes* (Japanese Medaka), Mouse, Rat, and Human [15–20]. Real time PCR, species specific primers can be found in Table 1, and non-real time PCR, species specific primers in Table 2.

2   Materials

2.1   DNA Isolation

2.1.1   C. elegans, Small Number

1. Worm Lysis Buffer: Standard 1× PCR Buffer in nuclease-free H_2O, 1 µg/mL Proteinase K.

2. Platinum worm pick.

3. Thin walled PCR tubes. Thermal cycler or heat block.

4. Dry ice and/or –80 °C freezer.

2.1.2   C. elegans, Large Numbers or Animal Tissue

1. K medium: 31.5 mM KCl, 51 mM NaCl in ddH_2O (*C. elegans*).

2. RNAlater Solution Optional for (animal tissue).

3. 15 mL conical tubes.

4. Liquid nitrogen.

Table 1
Real time PCR primers and conditions

Species	Genome	Target gene	Forward primer seq 5'–3'	Reverse Primer Seq 5'–3'	Amplicon (bps)	Annealing temperature (°C)	Reference
C. elegans	mt	nd-1	AGC GTC ATT TAT TGG GAA GAA GAC	AAG CTT GTG CTA ATC CCA TAA ATG T	75	62	[14]
	nuc	Cox-4	GCC GAC TGG AAG AAC TTG TC	GCG GAG ATC ACC TTC CAG TA	164	62	This chapter
H. sapiens	mt	tRNA-Leu(UUR)	CAC CCA AGA ACA GGG TTT GT	TGG CCA TGG GTA TGT TGT TA	107	62	[21]
	nuc	B2-microglobulin	TGC TGT CTC CAT GTT TGA TGT ATC T	TCT CTG CTC CCC ACC TCT AAG T	86	62	

Table 2
Quantitative, non-real time PCR primers and conditions

Species	Genome	Forward Primer Seq 5'–3'	Reverse Primer Seq 5'–3'	Amplicon (bps)	Annealing[a] temperature (°C)	Cycle[a] Number	Reference
Mouse	mt	CCC AGC TAC TAC CAT CAT TCA AGT	GAT GGT TTG GGA GAT TGG TTG ATG T	117	60	18	[17]
Rat	mt	CCT CCC ATT CAT TAT CGC CGC CCT TGC	GTC TGG GTC TCC TAG TAG GTC TGG GAA	211	60	21	[17]
Human	mt	CCC CAC AAA CCC CAT TAC TAA ACC CA	TTT CAT CAT GCG GAG ATG TTG GAT GG	221	60	18	[17]
D. melanogaster	mt	GCT CCT GAT ATA GCA TTC CCA CGA	CAT GAG CAA TTC CAG CGG ATA AA	151	61	19	[16]
	nuc	CGA GGG ATA CCT GTG AGC AGC TT	GTC ACT TCT TGT GCT GCC ATC GT	152	65	24	
D. rerio	mt	CAA ACA CAA GCC TCG CCT GTT TAC	CAC TGA CTT GAT GGG GGA GAC AGT	198	62	21	[16]
	nuc	ATG GGC TGG GCG ATA AAA TTG G	ACA TGT GCA TGT CGC TCC CAA A	233	60	27	
C. elegans	mt	CAC ACC GGT GAG GTC TTT GGT TC	TGT CCT CAA GGC TAC CAC CTT CTT CA	195	63	18	[16, 18]
	nuc	TCC CGT CTA TTG CAG GTC TTT CCA	GAC GCG CAC GAT ATC TCG ATT TTC	225	63	23	[18, 19]
O. latipes	mt	AAC TCC AAG TAG CAG CTA TGC AC	GAG GGG TAG AAG GCT TAC AAA AA	184	59	22	This chapter
	nuc	CTC ACA AAC ATC TTT GCA CTC AG	AGA ACC TCT CTC CAA AAC AIT CC	140	57	26	

F. grandis	mt	TTT ACA CAT GCA AGT ATC CG	CCG AAG GCT ATC AAC TTG AG	206	55	25	This chapter
	nuc	GCC GCT GCC TTC ATT GCT GT	ATG AGC TGG GTG TGC GCT GA	234	62	25	
F. heteroclitus	mt	ATC TGC ATG GCC AAC GCC TA	GGC GGT GCC AGT TTC CTT TT	264	62	24	[16, 20]
	nuc	GCC GCT GCC TTC ATT GCT GT	ATG AGC TGG GTG TGC GCT GA	234	62	24	

a Annealing temperature and cycle number are suggested starting points and may need optimization based on differences in laboratory equipment and PCR kit used

5. Dry ice.

6. Mortar and pestle (*C. elegans* and tough animal tissue such as muscle).

7. Handheld homogenizer (softer animal tissue such as liver).

8. Qiagen G/20 Genomic Tips Kit.

9. Isopropanol.

10. 70 % ethanol.

11. Glass Pasteur pipettes.

12. 1.7 mL microcentrifuge tubes.

13. 50 °C water bath.

14. Refrigerated microcentrifuge.

15. Tabletop centrifuge with 15 mL conical tube buckets.

2.1.3 Cultured Cells

1. Either the Qiagen 20/G Genomic Tips Kit and associated buffers (see above), or;

2. A QIAcube for automated DNA isolation with the QIAmp DNA Mini Kit for human samples or the DNeasy Blood and Tissue Kit (Qiagen) for animal samples [12].

3. Pellets of approximately 1×10^6 cells.

2.2 DNA Quantification (See Note 1)

1. PicoGreen dsDNA quantification reagent.

2. Lambda/HindIII DNA standard curve.

3. 1× TE buffer: 10 mM Tris–HCL pH 8.0, 1 mM EDTA.

4. Fluorescent plate reader with excitation filter at 480 nm and an emission filter at 520 nm (485 nm and 528 nm also work well).

5. Black or white bottom 96-well plate.

2.3 Real Time PCR

1. SYBR Green PCR Master Mix.

2. Standard 96-well PCR plate with optically clear sealing film.

3. Real Time PCR System (ABI 7300).

4. ABI Prism 7300 Sequence Detection Software.

5. Primers, species and target genome specific, *see* Table 1.

6. Nuclease-free H_2O.

2.4 Quantitative, Non-Real Time, PCR

1. Standard thermal cycler.

2. KAPA Long Range Hot Start DNA Polymerase Kit (KAPA Biosystems) (optimized for Human samples, *see* **Note 2**), or:

3. GoTaq Flexi PCR Kit (Promega), (optimized for *F. grandis* samples, *see* **Note 2**).

4. 0.2 mL PCR tubes.

5. PCR hood with germicidal lamp for sterilization.

6. Primers (species and target genome specific), *see* Table 2.

7. All materials from Subheading 2.2 DNA Quantification.

8. 0.1 mg/mL bovine serum albumin in nuclease-free H_2O.

9. 10 mM dNTPs Mix.

10. Nuclease-free H_2O.

11. Dedicated pipettes and sterile aerosol pipette tips for QPCR set up.

12. Different set of pipettes and regular tips for post-PCR analysis.

13. Distinct workstations for setting up and post-PCR analysis (*see* **Note 3**).

3 Methods

3.1 DNA Isolation

3.1.1 C. elegans, Small Number of Worms

1. Using a platinum worm pick, transfer six individual, L4 stage or later *C. elegans* into 90 μL of 1× worm lysis buffer pre-aliquoted into thin walled PCR tubes (*see* **Notes 4** and **5**), and freeze on dry ice (or at –80 °C) immediately. If using dry ice, once all samples are picked transfer to –80 °C for at least 10 min. This is usually done in triplicate for each sample and data are averaged (*see* **Note 6**).

2. Thaw samples, vortex briefly, and spin to collect contents at the bottom of the tube. In a standard thermal cycler or heat block, heat to 65 °C for 1 h, followed by 95 °C for 15 min, and then hold at 8 °C. This crude worm lysate will be used as template DNA for the real time PCR reactions and *does not need to be quantified*. This lysate can also be used for the non-real time quantitative PCR, if real time PCR is not available.

3.1.2 C. elegans, Large Number of Worms [15] or Animal Tissue (Skip **Steps 1** and 2)

1. Wash worms off of bacterial plate with K medium into a 15 mL conical tube, pellet at 2,200×*g* for 2 min, remove medium, and resuspend in 10 mL fresh medium. Gently rock tubes for 20 min to allow worms to clear gut contents. Pellet at 2,200×*g* for 2 min and wash 2× with fresh medium.

2. Resuspend worm pellet in a small volume of medium (residual medium left after wash) with a glass pipette and drip worm suspension directly into liquid nitrogen.

3. Snap frozen in liquid Nitrogen or placed in RNAlater solution and stored at –80 °C. As a guide, roughly 10–15 mg of *F. grandis* liver tissue is sufficient for DNA isolation.

4. Grind frozen worm pellets or tough tissue samples to a fine powder in a liquid nitrogen cooled mortar and pestle (*see* **Note 7**). A squeaking sound is heard when worms are sufficiently ground. Alternatively, if the tissue is not tough (i.e., liver tissue), it can

be manually homogenized in pre-aliquoted buffer G2 with RNAse A.

5. Scoop the powder into pre-aliquoted buffer G2 with RNAse A, as per the Qiagen 20/G Genomic Tips Handbook tissue protocol.

6. Follow the Qiagen 20/G Genomic Tips Tissue protocol for DNA isolation.

3.1.3 Cell Culture Samples [15]

1. Standard DNA isolation methods can be used. We routinely use the Qiagen 20/G Genomic Tips Kit or, for automated DNA isolation, the QIAcube with the QIAamp DNA Mini kit or DNeasy Blood and Tissue Kit can be used (*see* **Note 8**).

3.2 DNA Quantification [15]

1. DNA from large scale worm preparations, cultured cells or animal tissue needs to be quantified prior to real time or quantitative non-real time PCR. DNA from the small-scale worm lysis protocol does not.

2. Prepare a DNA concentration standard curve by diluting Lambda/HindIII DNA to 150, 100, 50, 25, 12.5, and 0 ng/µL in TE buffer (*see* **Note 9**).

3. Dilute DNA samples 1:10 in 1× TE buffer (*see* **Note 10**).

4. Add 5 µL of DNA and 95 µL of 1× TE buffer into two duplicate wells of a black or white 96-well plate (suitable for fluorescence measurements).

5. Add 5 µL of each Lambda/HindIII standard and 95 µL 1× TE into two duplicate wells.

6. The following three steps should be done in low light conditions (*see* **Note 11**). Prepare PicoGreen working solution (100 µL of working solution needed per well) by adding 5 µL PicoGreen reagent per 1 mL TE buffer (*see* **Note 12**).

7. Add 100 µL PicoGreen working solution to each well and incubate at room temperature in the dark for 10 min.

8. Measure the fluorescence of each sample with excitation at 480 nm and emission at 520 nm.

9. Determine DNA sample concentrations by comparing fluorescence values to those of the standard curve. If the DNA concentrations are far above the range of the curve, re-dilute the DNA and measure again (*see* **Note 13**).

10. Dilute the sample to 3 ng/µL in TE buffer.

3.3   Real Time PCR

3.3.1   C. elegans Samples with Standard Curve

1. Prepare the standard curve (*see* **Note 14**) as follows: Thaw an aliquot of the mtDNA copy number standard curve plasmid (50,000 copies/µL) and dilute to 32,000 copies/µL. Serially dilute 1:1 down to 4,000 copies/µL [14]. Add 2 µL of each dilution and a 0 copies/µL control (TE buffer or water) to separate wells in the 96-well PCR plate to be used in the

copy number PCR. A typical standard curve contains 64,000, 32,000, 16,000, 8,000, 4,000, and 0 copies per well.

2. Proceed with real time PCR setup for all other samples.

3.3.2 All Other Real Time PCR Samples, Without Standard Curve [21]

1. Assemble the PCR reactions as follows: combine 2 µL of template DNA (standard curve dilution, worm lysate, or 3 ng/µL isolated DNA), 2 µL of mtDNA target specific primer pair (400nM final concentration each, *see* **Note 15**), 12.5 µL SYBR Green PCR Master Mix, and 8.5 µL H₂O in 1 well of the 96-well PCR plate. Each sample is amplified in triplicate and the data are averaged (*see* **Note 16**). Repeat this step in separate wells using nuclear DNA specific primers.

2. When analyzing a large number of samples with the same primer pair, a master mix is made containing the reagents that are common to all reactions (SYBR Green Master Mix, H₂O, and primers) and aliquoted into individual reactions.

3. Cycle in an ABI 7300 Real Time PCR System (or comparable system) as follows: 50 °C for 2 min, 95 °C for 10 min, 40 cycles of 95 °C for 15 s and annealing temperature (primer specific, Table 1) for 60 s. A dissociation curve is also calculated for each sample to ensure presence of a single PCR product.

3.3.3 Quantitative (Non-Real Time) PCR

1. For samples from *C. elegans*, *D. melanogaster*, *F. heteroclitus*, *F. grandis*, *O. latipes*, *D. rerio*, Mouse, Rat, and Human, relative mtDNA content can be measured in a quantitative, non-real time, PCR reaction in which the PCR product is quantified after completion of the reaction. Nuclear copy number can also be measured using this protocol, and primers are listed in Table 2 for some nuclear targets; however, this is not often necessary (*see* **Note 17**).

2. Extra control reactions are necessary for this protocol. Be sure to include a "50 % control" that contains control template DNA (or worm lysate) diluted 1:1 with H₂O or TE buffer prior to being added to the reaction (it is not advised to simply add ½ the volume of the control sample), and a "No Template control" that contains only H₂O or TE buffer in place of template DNA.

3. Initially, it is vital to determine the appropriate cycle number for the reaction (*see* **Note 18**). Listed in Table 2 are approximate cycle numbers that will serve as good starting points for cycle number optimization. The cycle number is correct when the "50 % control" reaction results in 40–60 % of the PCR product of the undiluted control reaction.

4. Specific reaction conditions for two PCR kits are presented below. The first uses the KAPA Biosystems LongRange Hot Start PCR kit, and has been optimized with Human samples.

The second used the GoTaq Flexi Kit from Promega and has been optimized for *F. grandis* samples (*see* **Note 2** for more detailed information).

5. Using the KAPA LongRange Hot Start kit, the reactions are prepared as follows:

   (a) A master mix is made if several samples are being run simultaneously, which consists of the following components added in this order:

   - 24.5 µL nuclease-free H_2O (for a final volume of 50 µL).

   - 10 µL of 5× buffer solution (vortex at this stage).

   - 1 µL of BSA in nuclease-free H_2O (0.1 mg/mL stock, 2 ng/µL final).

   - 1 µL of dNTPs (10 mM stock, 200 µM final).

   - 2.5 µL of each primer working solution (10 µM stock, 0.5 µM final, except *O. latipes*, 7.5 µM stock).

   - 3.5 µL of $MgCl_2$ (25 mM stock, 1.75 mM final).

   - 0.5 µL KAPA LongRange Hot Start DNA Polymerase (2.5U/µL).

   - Vortex and spin.

   (b) Aliquot the master mix into the appropriate number of tubes, and add 15 ng purified DNA (5 µL if diluted to 3 ng/µL) or 5 µL of worm lysate as template to the pre-aliquoted master mix. Also, add 5 µL of "50 % control" and "No Template control" to the appropriate tubes.

   (c) The PCR amplification profile is as follows: 94 °C for 3 min, followed by the optimized number of cycles of 94 °C for 15 s, annealing temperature (Table 2) for 45 s, and 72 °C for 45 s. To complete the profile perform a final extension for 10 min at 72 °C.

6. Using the GoTaq Flexi Kit, the reactions are assembled as follows:

   (a) A master mix is made if several samples are being run simultaneously, which consists of the following components added in this order:

   - 9.5 µL nuclease-free water (for a final volume of 25 µL).

   - 5 µL of 5× Colorless GoTaq Flexi Buffer (vortex at this stage).

   - 1 µL of PCR Nucleotide mix (10 mM stock, 400 µM final).

   - 1 µL of each primer solution (10 µM stock, 0.4 µM final).

- 1 μL MgCl$_2$ (25 mM stock, 1 mM final).
- 0.5 μL GoTaq DNA polymerase (5U/μL)
- Vortex and spin.

(b) Aliquot the master mix into the appropriate number of tubes, and add 60 ng purified DNA (6 μL if diluted to 10 ng/μL) as template to the pre-aliquoted master mix.

(c) The PCR amplification profile is as follows: 94 °C for 2 min, followed by the optimized number of cycles of 94 °C for 30 s, annealing temperature (Table 2) for 30 s, and 72 °C for 1 min. To complete the profile perform a final extension for 5 min at 72 °C.

7. The primers provided in Table 2 have been tested and verified to result in a single, specific PCR product; however, when first optimizing the assay it is recommended to check the specificity of the PCR products on an agarose gel, as conditions may vary slightly based on laboratory equipment and PCR kits used. It is critical to obtain a single product to ensure accurate quantification in **step 8**.

8. Quantify the resulting PCR products (similar to Subheading 3.2). Add 10 μL of each PCR product and 90 μL of TE buffer to each of two duplicate wells of a white or black 96-well plate. Also, add 10 μL of the DNA concentration standard curve (*see* Subheading 3.2) and 90 μL of TE buffer to each of two duplicate wells of the same plate.

9. Follow **steps 6–8** of Subheading 3.2 for fluorescent quantification of DNA. The standards are not used to calculate DNA concentration in this protocol, but are useful to assure that the samples fall within the linear range of the instrument and can be helpful when comparing samples on different plates.

3.4 Data Analysis

3.4.1 C. elegans Samples, Real Time PCR with Standard Curve

1. Obtain mtDNA cycle threshold (*Ct*) values from the Real Time PCR software, and average the *Ct* values for the triplicate reactions. If any of the triplicate *Ct* values vary by more than 0.5 U from the others they should be removed prior to analysis.

2. Obtain the *Ct* values for the standard curve reactions and perform a logarithmic regression with *Ct* on the Y-axis and copy number on the *X*-axis (*see* **Note 19**).

3. Compare the *Ct* values of the samples with those of the standard curve to determine the copy number per PCR reaction. We use the following equation:

(a) $e^{[(\text{Sample } Ct - \text{slope})/y - \text{intercept}]} = \text{mtDNA copy number per reaction}$

4. Assuming worms were picked at a ratio of 1 worm per 15 and 2 μL were used as template, multiply this number by 7.5 to get mtDNA copy number per worm (*see* **Note 20**).

3.4.2 H. sapiens Samples, Real Time PCR Without Standard Curve [15]

1. Obtain both mtDNA and nucDNA *Ct* values from Real Time PCR software and average the *Ct* values from triplicate reactions.

2. To determine the mitochondrial DNA content relative to nuclear DNA, use the following equations:

   (a) $\Delta Ct = (\text{nucDNA Ct} - \text{mtDNA Ct})$

   (b) Relative mitochondrial DNA content $= 2 \times 2 \Delta CT$

3.4.3 Quantitative, Non-Real Time PCR

1. Obtain the fluorescence values from the plate reader software.

2. Subtract the "No Template Control" values from all other PCR product values (Not standard curve values).

3. Average the values of the duplicate wells.

4. Assure that the "50 % control" values fall between 40 and 60 % of the undiluted control values.

5. The fluorescence values can be used directly to compare relative mitochondrial DNA content between samples.

4 Notes

1. DNA from small scale worm lysis does not need to be quantified prior to PCR; however, quantification is required for purified DNA from large scale preparations.

2. The GeneAmp XL PCR kit from Applied Biosystems that was previously used for the QPCR assay has been discontinued. We have now optimized the QPCR assay using KAPA LongRange Hot Start DNA Polymerase with Human samples and the GoTaq Flexi PCR kit with *F. grandis* samples. We expect these protocols to be easily adaptable to the other species that we provide primers for. It is also likely that other PCR kits will work with this protocol, though some optimization may be required (discussed in further detail in Subheading 3.4 and **Note 17**).

3. Maintaining distinct pre- and post-PCR workstations and pipettes helps to reduce the possibility of contamination of the pre-PCR workstation, and therefore new PCR reactions, with PCR product. This is especially important when the same PCR target will be amplified repeatedly from many different samples, as is often the case with experiments that use these protocols. We use a PCR hood equipped with a UV sterilizing lamp for reaction assembly, and completed reactions are never opened

in this room. This is not a concern with real time PCR as samples are not processed post-PCR.

4. Six worms in 90 µL is the standard condition we routinely use; however, different numbers of worms can be used. For worms at or past the L4 stage, we use a ratio of one worm per 15 µL lysis buffer, and for younger worms, we pick one worm per 10 µL buffer.

5. We culture worms on K-media plates, as opposed to the standard NGM, because K-media supports thinner bacterial lawns, resulting in less transfer of bacteria when picking worms for PCR.

6. Samples are picked in triplicate and each sample is amplified in triplicate, resulting in nine individual real time PCR reactions per data point. For example, if comparing ethidium bromide treated worms with controls, we would pick three tubes, each with six worms in 90 µL lysis buffer, from the treated group (EtBr 1, 2, and 3) and the same from the control group (C1, 2, and 3). After lysis, each of these samples would be amplified via real time PCR in triplicate, resulting in nine PCR reactions for each treatment group.

7. When grinding frozen worm pellets, we typically pack the outside of the mortar with dry ice, and then chill the mortar and pestle with a small amount of liquid nitrogen. Once the liquid nitrogen has boiled off the worm pellets are added to the mortar and ground. Care should be taken in the initial few "grinds" as the larger pellets have a tendency to "jump" out of the mortar.

8. We routinely linearize the Human mitochondrial DNA from QIAcube preparations by digesting with the PvuII restriction enzyme prior to analysis [15]. This step is performed for an unrelated protocol, however, we measure copy number on these linearized templates. While we do not expect this digestion to be necessary for copy number analysis, this has not been exhaustively tested.

9. We prepare larger stocks of the standard curve dilutions that are routinely reused to reduce variability. Store at 4 °C.

10. Depending on the number of worms or cells or the amount of tissue used for DNA extraction, this dilution will change. The aim is to get the DNA concentration in the range of the standard curve for an accurate measurement of concentration.

11. Overhead lab lights are turned off and shades are pulled down. During the 10 min incubation, the plates are covered with aluminum foil or placed in a drawer.

12. PicoGreen working solution should be prepared in excess of what is needed to account for loss during pipetting. We pour the

working solution into a reservoir and add 100 μL to each well of the plate using an 8-channel micropipette, and typically make between 500 μL and 1,000 μL excess per full 96-well plate.

13. As stated in **Note 8**, the aim is to get the DNA concentrations in the range of the standard curve. If the DNA concentration is too high, further dilute and measure again. If too low, the undiluted samples can be measured.

14. The standard curve for *C. elegans* copy number allows us to calculate *actual* mtDNA copy number per worm. Most methods can only determine copy number on a per nuclear copy basis. The standard curve plasmid is pCR2.1 with a 75 bp fragment of the nd-1 gene, containing the real time PCR primer target sequence, cloned into the multiple cloning site [14]. Based on the molecular weight of the plasmid (MW = 4,006) and its concentration, we can accurately determine the number of plasmids (and therefore the number of nd-1 primer targets) per μL. We store this plasmid at a concentration of 100,000 copies per μL in single use aliquots (to avoid freeze thaw cycles) at −20 °C. We routinely make duplicate serial dilutions of the standard curve and include both sets of dilutions in our real time PCR reactions to ensure accuracy.

15. Individual primers stocks (100 μM) are kept at −20 °C, and working dilutions (25 μM) at 4 °C. When assembling PCR reactions, working dilutions are mixed 1:1 and further diluted to 5 μM each. 2 μL of this primer mix is added to each PCR reaction, resulting in final primer concentrations of 400 nM each.

16. As stated in **Note 4**, each sample is analyzed in three separate real time PCR reactions, and the results of these three reactions are averaged.

17. In the quantitative, non-real time PCR for all samples other than *C. elegans*, the concentration of total DNA is known. Total DNA concentration is based almost entirely on nuclear DNA, and the same amount of total DNA is added to each PCR reaction. For this reason, we effectively start each PCR reaction with the same number of nuclear DNA copies, and the values obtained for relative mitochondrial DNA content do not need to be normalized prior to comparison between samples.

18. PCR product is created proportionally to the amount of specific target sequence in the template during the exponential phase of the reaction. Therefore, for quantitative results, the reaction must be stopped while in this exponential phase. Assuring that the "50 % control" sample results in 40–60 % of the PCR product of undiluted control will also assure that the reaction is in the exponential phase [12, 13].

19. When plotting the standard curve *Ct* values, assure that all values fall on the regression line. If a single sample is not on the

line, it can be removed. Also, visually inspect the *Ct* values for accuracy. Each time the starting DNA concentration is reduced by half, the *Ct* should increase by 1.

20. If we pick six worms in 90 µL of lysis buffer, we assume that 15 µL of this lysate is roughly equal to one worm, therefore copy number per 15 µL = copy number per worm. This calculation can be adjusted based on the ratio of worms to volume of lysis buffer.

Acknowledgements

We thank Aleksandra Trifunovic for sharing the mtDNA copy number plasmid for *C. elegans* with us, and we thank Kevin Kwok for supplying us with medaka tissue. This work was supported by P42 ES010356-10A2, and the National Science Foundation (NSF) and the Environmental Protection Agency under NSF Cooperative Agreement EF-0830093, Center for the Environmental Implications of NanoTechnology (CEINT). Any opinions, findings, conclusions, or recommendations expressed in this material are those of the author(s) and do not necessarily reflect the views of the NSF or the EPA. This work has not been subjected to EPA review and no official endorsement should be inferred. The funding sources had no role in experimental design, data collection, or interpretation.

References

1. Wallace DC (2010) Mitochondrial DNA mutations in disease and aging. Environ Mol Mutagen 51(5):440–450. doi:10.1002/em.20586

2. Copeland WC (2010) The mitochondrial DNA polymerase in health and disease. Subcell Biochem 50:211–222. doi:10.1007/978-90-481-3471-7_11

3. Campbell CT, Kolesar JE, Kaufman BA (2012) Mitochondrial transcription factor A regulates mitochondrial transcription initiation, DNA packaging, and genome copy number. Biochim Biophys Acta 1819(9–10):921–929. doi:10.1016/j.bbagrm.2012.03.002

4. Copeland WC (2012) Defects in mitochondrial DNA replication and human disease. Crit Rev Biochem Mol Biol 47(1):64–74. doi:10.3109/10409238.2011.632763

5. Suomalainen A, Isohanni P (2010) Mitochondrial DNA depletion syndromes–many genes, common mechanisms. Neuromuscul Disord 20(7):429–437. doi:10.1016/j.nmd.2010.03.017

6. Copeland WC (2008) Inherited mitochondrial diseases of DNA replication. Annu Rev Med 59:131–146. doi:10.1146/annurev.med.59.053006.104646

7. Rolo AP, Palmeira CM (2006) Diabetes and mitochondrial function: role of hyperglycemia and oxidative stress. Toxicol Appl Pharmacol 212(2):167–178. doi:10.1016/j.taap.2006.01.003

8. Yu M (2011) Generation, function and diagnostic value of mitochondrial DNA copy number alterations in human cancers. Life Sci 89(3–4):65–71. doi:10.1016/j.lfs.2011.05.010

9. Coskun P, Wyrembak J, Schriner SE, Chen HW, Marciniack C, Laferla F, Wallace DC (2012) A mitochondrial etiology of Alzheimer

and Parkinson disease. Biochim Biophys Acta 1820(5):553–564. doi:10.1016/j.bbagen.2011.08.008

10. Blanche S, Tardieu M, Rustin P, Slama A, Barret B, Firtion G, Ciraru-Vigneron N, Lacroix C, Rouzioux C, Mandelbrot L, Desguerre I, Rotig A, Mayaux MJ, Delfraissy JF (1999) Persistent mitochondrial dysfunction and perinatal exposure to antiretroviral nucleoside analogues. Lancet 354(9184):1084–1089. doi:10.1016/S0140-6736(99)07219-0, S0140-6736(99)07219-0 [pii]

11. Fetterman JL, Pompilius M, Westbrook DG, Uyeminami D, Brown J, Pinkerton KE, Ballinger SW (2013) Developmental exposure to second-hand smoke increases adult atherogenesis and alters mitochondrial dna copy number and deletions in apoE(-/-) mice. PLoS One 8(6):e66835. doi:10.1371/journal.pone.0066835

12. Pavanello S, Dioni L, Hoxha M, Fedeli U, Mielzynska-Svach D, Baccarelli AA (2013) Mitochondrial DNA copy number and exposure to polycyclic aromatic hydrocarbons. Cancer Epidemiol Biomarkers Prev 22:1722–1729. doi:10.1158/1055-9965.EPI-13-0118

13. Janssen BG, Munters E, Pieters N, Smeets K, Cox B, Cuypers A, Fierens F, Penders J, Vangronsveld J, Gyselaers W, Nawrot TS (2012) Placental mitochondrial DNA content and particulate air pollution during in utero life. Environ Health Perspect 120(9):1346–1352. doi:10.1289/ehp.1104458

14. Bratic I, Hench J, Henriksson J, Antebi A, Burglin TR, Trifunovic A (2009) Mitochondrial DNA level, but not active replicase, is essential for Caenorhabditis elegans development. Nucleic Acids Res 37(6):1817–1828. doi:10.1093/nar/gkp018

15. Furda AM, Bess AS, Meyer JN, Van Houten B (2012) Analysis of DNA damage and repair in nuclear and mitochondrial DNA of animal cells using quantitative PCR. Methods Mol Biol 920:111–132. doi:10.1007/978-1-61779-998-3_9

16. Hunter SE, Jung D, Di Giulio RT, Meyer JN (2010) The QPCR assay for analysis of mitochondrial DNA damage, repair, and relative copy number. Methods 51(4):444–451. doi:10.1016/j.ymeth.2010.01.033

17. Ayala-Torres S, Chen Y, Svoboda T, Rosenblatt J, Van Houten B (2000) Analysis of gene-specific DNA damage and repair using quantitative polymerase chain reaction. Methods 22(2):135–147. doi:10.1006/meth.2000.1054, S1046-2023(00)91054-5 [pii]

18. Meyer JN, Boyd WA, Azzam GA, Haugen AC, Freedman JH, Van Houten B (2007) Decline of nucleotide excision repair capacity in aging Caenorhabditis elegans. Genome Biol 8(5):R70. doi:10.1186/gb-2007-8-5-r70

19. Boyd WA, Crocker TL, Rodriguez AM, Leung MC, Lehmann DW, Freedman JH, Van Houten B, Meyer JN (2010) Nucleotide excision repair genes are expressed at low levels and are not detectably inducible in Caenorhabditis elegans somatic tissues, but their function is required for normal adult life after UVC exposure. Mutat Res 683(1–2):57–67. doi:10.1016/j.mrfmmm.2009.10.008

20. Jung D, Cho Y, Meyer JN, Di Giulio RT (2009) The long amplicon quantitative PCR for DNA damage assay as a sensitive method of assessing DNA damage in the environmental model, Atlantic killifish (Fundulus heteroclitus). Comp Biochem Physiol C Toxicol Pharmacol 149(2):182–186. doi:10.1016/j.cbpc.2008.07.007, S1532-0456(08)00139-7 [pii]

21. Venegas V, Halberg MC (2012) Measurement of mitochondrial DNA copy number. Methods Mol Biol 837:327–335. doi:10.1007/978-1-61779-504-6_22

Chapter 4

NAD⁺ Content and Its Role in Mitochondria

Wei Li and Anthony A. Sauve

Abstract

Nicotinamide adenine dinucleotide (NAD⁺) is a central metabolic coenzyme/cosubstrate involved in cellular energy metabolism and energy production. It can readily be reduced by two electron equivalents and forms the NADH form, which is the minority species to NAD⁺ under most physiologic conditions. NAD⁺ plays an important role in not only oxidation–reduction reactions in cells but also as a signaling molecule. For example, NAD⁺ plays a key role in mitochondrial function via participation in pyruvate dehydrogenase, tricarboxylic acid cycle, and oxidative phosphorylation chemistries. It also serves as a substrate for deacylases SIRT3, SIRT4, and SIRT5, which modify protein posttranslational modifications on lysine within the mitochondrial compartment. Recent work has highlighted the biological significance of dynamic changes to mitochondrial NAD⁺. This has increased the need for standardized and effective methods to measure NAD⁺ contents in this organelle. To determine NAD⁺ concentrations in cells, and specifically in mitochondria, we describe two assays for NAD⁺ determinations: An Enzymatic Cycling Assay and Isotope Dilution. The cycling assay contains sample NAD⁺, lactate, lactate dehydrogenase, diaphorase, and resazurin. The isotope dilution assay uses synthetic ^{18}O-NAD⁺ as an internal standard, and treated samples are fractionated by HPLC and then NAD⁺ concentration determined by the ^{16}O- and ^{18}O-NAD⁺ peak (664/666) ratio in positive mode MS.

Key words NAD⁺, Mitochondria, Energy metabolism, Oxidative phosphorylation, NAD⁺ measurement, NAD⁺-NADH cycling assay, Lactate, Lactate dehydrogenase (LDH), Diaphorase, Resazurin, Fluorescence, Plate reader, ^{18}O-NAD⁺, Isotope dilution, C-18 column

1 Introduction

Nicotinamide adenine dinucleotide (NAD⁺, Fig. 1) is a metabolite derived from vitamin B3 (nicotinamide and nicotinic acid) or tryptophan [1]. NAD⁺ is centrally incorporated into energy metabolism and forms NADH by reduction with a hydride equivalent (H–). The dinucleotides are also found in phosphorylated forms (NADP/NADPH), which behave analogously as hydride-accepting and hydride-donating metabolites within cells, although these redox pairs are kept in chemical opposition. For example, NAD⁺ is largely maintained in oxidized form, whereas NADP⁺ is predominantly found in its reduced form NADPH. Because NAD⁺ plays important

Carlos M. Palmeira and Anabela P. Rolo (eds.), *Mitochondrial Regulation*, Methods in Molecular Biology, vol. 1241, DOI 10.1007/978-1-4939-1875-1_4, © Springer Science+Business Media New York 2015

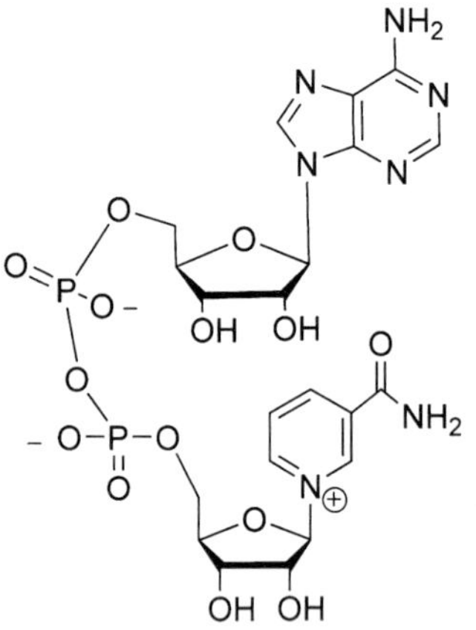

Fig. 1 The structure of NAD⁺

roles in energy metabolism and oxidation process its homeostasis in cells is tightly maintained and its redox balance is also critically maintained. For example, NADH is anaerobically converted back to NAD⁺ by pyruvate reduction to form lactate. NADH is also reacted back to NAD⁺ via reaction with Complex I, the NADH/coenzyme Q reductase of the mitochondrial electron transport chain [2]. NADH content in cells is typically maintained below or at 10 % of total cellular NAD⁺ content [3], with larger amounts associated with metabolic disruption. Increasing interest in how NAD⁺ levels are regulated include concerns about how NAD⁺ contents affect net energy and metabolic fluxes in cells [4, 5]. In addition, key regulatory steps in NAD⁺ production and turnover are sensitive to cellular stresses [6, 7], highlighting the potential for changes in NAD⁺ concentration to operate as a driver for metabolic and physiologic adaptation.

A previously underappreciated role of cellular NAD⁺ is through regulation of the activities of key signaling enzymes, such as PARPs and sirtuins [6]. These enzymes appear to respond dynamically to cellular changes in NAD⁺ concentrations, in combination with other cellular stimuli. The powerful effects of these signaling components in regulating key and central metabolic processes, such as fuel switching, cell apoptosis, proliferative signaling, and mitochondrial biogenesis, have renewed interest in NAD⁺ metabolism in general [8], and focused attention on several finer points of this metabolism, which may have regulatory significance. Among key considerations is how compartmental NAD⁺ levels change in response to pharmacologic or physiologic perturbations. Three major cellular depots of NAD⁺ have been identified, namely, nuclear, cytosolic, and mitochondrial.

The methods to determine NAD⁺ content in mitochondria are similar to those used to determine NAD⁺ contents in other cellular

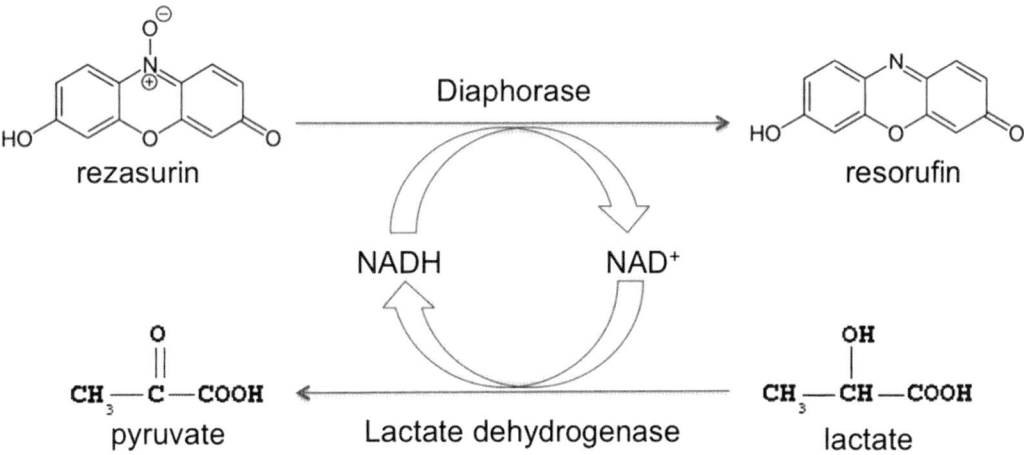

Fig. 2 Plate reader NAD+ cycling assay requiring lactate dehydrogenase and diaphorase

compartments, although a key aspect of this organelle is the need to keep the compartment intact through the cell fractionation process. Mitochondria are membrane-bound organelles found in most eukaryotic cells. Previous research indicates that total NAD+ in cells can have significant mitochondrial components, and that mitochondrial NAD+ does not readily leak across the inner mitochondrial membrane [7, 9]. These considerations are key to being able to reliably measure NAD+ concentrations in this organelle. Our laboratory has developed methods for measurement of NAD+ concentrations in mitochondria, and data obtained via these methods support the concept that mitochondrial NAD+ content is dynamically responsive to physiologic perturbations as well as pharmacologic interventions.

NAD+ contents in cells or mitochondria can be determined by a known cycling assay [10–12]. Briefly, after homogenization of sample in perchloric acid, and neutralization, samples are assayed in a cycling reaction combining L-lactate, lactate dehydrogenase (LDH), diaphorase, and the dye resazurin. In an initial reaction, NAD+ reacts with lactate catalyzed by LDH to form NADH (Fig. 2), which then binds to the enzyme diaphorase to reduce resazurin to form the reduced dye resorufin, regenerating NAD+ (Fig. 2). Resorufin formation is monitored versus time by fluorescence (530 nm excitation, 580 nm emission) by a plate reader. Standards of known NAD+ concentration are prepared and processed through similar procedures to generate a standard curve. Concentrations of NAD+ are expressed as pmol NAD+/10^6 cells or pmol NAD+/mg mitochondrial protein.

Another methodology to determine NAD+ concentration in cells or mitochondria is an isotope dilution approach. This method requires synthesis of isotopically labeled NAD+ for use as an internal standard for mass spectrometer measurements of NAD+ in biological samples. The synthesis of ^{18}O-NAD+ is described by our published work [7, 13]. Samples, as above, require homogenization with perchloric acid containing ^{18}O-NAD+. By high-performance liquid chromatography (HPLC), fractions containing NAD+ are separated, eluted, collected, dried, and analyzed by either electrospray ionization (ESI) or MALDI mass spectrometry (MS). The amount of sample NAD+ is obtained by the ratio of the labeled standard to the unlabeled metabolite (the ^{16}O- and ^{18}O-NAD+ peaks 664/666 ratio). This method offers the advantage of not requiring complete recovery of NAD+ to obtain accurate NAD+ quantitation, since it presumes all losses are proportionate to both internal standard as well as the metabolite in the sample. This is highly convenient for obtaining high accurate measurements of NAD+ content, even if samples experience instability, or if there are interfering factors in the mass spectrum measurement [13].

2   Materials

2.1   Buffers for Sample Neutralization

7 % Perchloric Acid.

2 M NaOH.

500 mM K_2HPO_4 (pH = 9).

2.2   Enzymes, Buffers for Cycling Assay

10× reaction buffer: 250 mM Tris, 50 mM $MgCl_2$, and 500 mM KCl. Store at 4 °C.

Resazurin stock solution (e.g., 27 mM). Aliquot and store at 4 °C. Avoid light.

Lactate dehydrogenase (LDH), purchased.

L-lactate: 22.5 mM solution, fresh made.

Diaphorase stock solution: Dissolve diaphorase into 250 mM Tris buffer (pH 7.5) to make a final concentration 0.35 U/μL. Store at 4 °C.

NAD+ stock solution (e.g., 7.5 μM) for standard curve preparation. Aliquot and store at −20 °C.

2.3   Other Assay Equipment/Supplies

Hemocytometry.

Vortex Mixer.

Sonicator (e.g., UltraSonic Cleanser).

pH paper.

Black flat bottom 96-well assay plate for fluorometry.

Multichannel pipette.

SpectraMax M5/M5 (Molecular Devices).

SoftMax Pro 5 Software (Molecular Devices, MDS Analytical Technologies).

2.4 Additional Materials for Isotope Assay	1 M Perchloric acid containing 1,750 nM ^{18}O-NAD$^+$ (95 % isotopic enrichment of ^{18}O-NAD$^+$). C-18 column. HPLC machine and software.

3 Methods

NAD$^+$ concentration can be measured in either cell pellets, isolated mitochondria or animal tissue samples. There are two different methodologies to determine NAD$^+$ concentration: by cycling assay or by isotope dilution. *See* Notes **1–4** for special details.

3.1 NAD$^+$ Concentration Determination by Cycling Assay

3.1.1 Sample Neutralization

1. For cell samples, count cells in 10^6 by hemocytometry. For animal tissue samples, grind tissue at low temperature (in liquid nitrogen) and weigh in mg (20–70 mg) before next step. For mitochondria, isolate mitochondria, and measure protein concentration. Total protein content in mitochondria should be at least 100 µg to provide solid measurements. Save lysate from last pelleting step for additional protein and NAD$^+$ measurements.

2. Treat cell pellets, tissue samples or isolated mitochondria with ice-cold 50–100 µL of 7 % perchloric acid.

3. Vortex for 30 s, sonicate for 5 min in ice bath. Then vortex again for 30 s, sonicate for 5 min. (Repeat for total of four cycles).

4. Spin down acid-treated samples to remove insolubles. Collect supernatants and transfer to new tubes.

5. Neutralize the supernatants with 2 M NaOH followed by 500 mM K_2HPO_4 (pH = 9) to pH = 7. Use pH paper to determine the pH of samples during neutralization. Do not overshoot pH. The phosphate buffer is designed to be used to neutralize to final pH. Repellet samples.

6. Neutralized samples can be used for the following cycling assay to determine NAD$^+$ concentration. Samples can also be stored at –20 or –80 °C for additional use.

3.1.2 NAD$^+$ Standard Curve Preparation

1. Prepare a fresh NAD$^+$ stock solution (e.g., 7.5 and 15 µM) in neutralized perchloric acid as described in the prior section. In a black 96-well plate, load different amounts of NAD$^+$ stock solution (e.g., 0, 1, 2, 4, 10, 10 µL (15 µM) with the balance of solution being neutralized perchloric acid to a final volume of 10 µL into wells to obtain final concentrations in 150 µL: 0 (blank), 50 nM, 100 nM, 200 nM, 500 nM, and 1 µM.

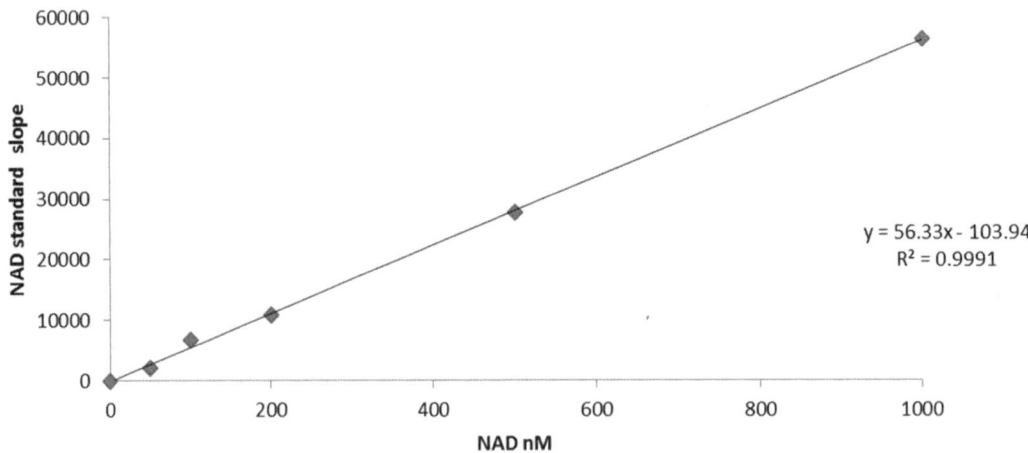

Fig. 3 Example NAD⁺ standard curve

2. Load cycling assay buffer (final concentration in 150 µL: 25 mM Tris pH 7.5, 5 mM MgCl$_2$, 50 mM KCl, 54 µM resazurin, 0.4 U/mL lactate dehydrogenase, and 2.25 mM L-lactate) onto the plate to make 130 µL per each well.

3. Add 20 µL of diluted diaphorase (0.035 U/µL in 25 mM Tris pH 7.5) with multichannel pipette to each well to initiate each reaction with final concentration of 0.7 U/well.

4. Immediately put the 96-well plate into a plate reader to determine formation of resorufin. (Molecular Devices SpectraMax M5/M5, SoftMax Pro 5 Software: 530 nm excitation, 580 nm emission). Record readings every 27 s for 15 min. Plot the fluorescence at each time point to obtain slopes.

5. Make a NAD⁺ standard curve by plotting slope value versus NAD⁺ concentration (Fig. 3). Make sure the whole system works well and the NAD⁺ standard curve looks good (see R^2 value) before measurement of biological samples.

3.1.3 Measurement of NAD⁺ Concentration by Cycling Assay

1. Spin down samples before use.

2. In a black 96-well plate, load 2 and 4 µL of neutralized samples onto each well and add 8 and 6 µL of cycling assay buffer to make the sample volume 10 µL per each well. For good statistics these should be run in duplicate and all samples used for determining concentration.

3. NAD⁺ standard curve is needed to prepare at the same plate for each run. For a NAD⁺ standard curve, dilute 7.5 µM NAD⁺ stock solution into wells to obtain final concentrations of 0 (blank), 50 nM, 100 nM, 200 nM, 500 nM, and 1 µM (15 µM).

4. Load 120 μL of cycling assay buffer (final concentration in 150 μL: 25 mM Tris pH 7.5, 5 mM $MgCl_2$, 50 mM KCl, 54 μM resazurin, 0.4 U/mL lactate dehydrogenase, and 2.25 mM L-lactate) onto the plate.

5. Add 20 μL of diluted diaphorase (0.035 U/μL in 25 mM Tris pH 7.5) with multichannel pipette to each well to initiate each reaction with final concentration of 0.7 U/well.

6. Immediately put the 96-well plate into a plate reader to determine formation of resorufin. (Molecular Devices SpectraMax M5/M5, SoftMax Pro 5 Software: 530 nm excitation, 580 nm emission). Record readings every 27 s for 15 min. Plot the fluorescence at each time point to obtain slopes.

7. Use the sample slopes to determine NAD⁺ concentration in each well by comparison to the standard curve.

8. Convert the NAD⁺ concentration calculated for each well to the original NAD⁺ concentration in samples by correction for dilutions. Calculate the final values in pmol NAD⁺/10^6 cells or pmol NAD⁺/mg protein or pmol NAD⁺/mg tissue.

3.1.4 Data Processing and Correction

1. SoftMax Pro 5 Software settings: Read Mode: Fluorescence, Top read; Wave lengths: Ex 530, Em 580; Sensitivity: Readings: 6, PMT: Medium; Timing: 15:00, Internal: 0:27, Reads: 34.

2. Obtain the slope value of each well from SoftMax Pro 5 Software.

3. Get the NAD⁺ concentration in each well by comparing the slopes to the NAD⁺ standard curve.

4. Then convert the NAD⁺ concentration in each well to the original NAD⁺ concentration in samples by correction for dilutions. For example, if 4 μL of sample is used for each well reaction of 150 μL, then the original sample NAD⁺ concentration (in pmol/μL) is: NAD⁺ conc. in well * 150/4 = 37.5 * NAD⁺ conc. in well.

5. Finally, by correction to the sample cell counts (in 10^6) or the sample tissue weight (in mg), or protein concentration, original NAD⁺ concentration is calculated in pmol/10^6 cells, pmol/mg protein, or pmol/mg tissue.

3.2 NAD⁺ Concentration Determination by Isotope Dilution

1. Seed cells in media and grow at 37 °C for 24–48 h. Remove media and treat cells with trypsin (0.5 mL, 0.05 %) for 5 min to detach them from the flask. Add fresh media and determine cell counts (in 10^6) by hemocytometry (with trypan blue staining). Gently pellet cells by centrifugation. For mitochondria, isolate intact mitochondria and measure protein content per unit volume. Total protein content in mitochondria should be at least 100 μg to provide solid measurements.

2. Add 400 µL ice-cold 1 M Perchloric Acid containing 700 pmol of ^{18}O-NAD^+ (95 % isotopic enrichment of ^{18}O-NAD^+, synthesis and assay of which is described in ref. 13) to the cell pellets or to isolated mitochondria.

3. Vortex for 30 s, sonicate for 5 min. Then vortex again for 30 s, sonicate for 5 min. (four times total, keep samples ice cold as much as possible).

4. Spin down acid-treated samples to remove insolubles. Collect supernatants and transfer to new tubes.

5. Neutralize the acidic supernatant with 5.5 M NaOH (65 µL) and 1 M K_2HPO_4 (pH = 7.5) to pH = 7 as determined by pH paper. Repellet solids by a final centrifugation.

6. Chromatograph supernatant (100 µL sample) on a C-18 column (Solid phase: EC 250/4.6 NUCLEOSIL 100-3 C18 HD column; Mobile phase: 20 mM ammonium acetate; 1 mL/min).

7. Collect the peak containing NAD^+ as determined by the elution position of an NAD^+ standard.

8. Lyophilize eluant and redissolve the residue in 10 µL water then co-spot with CHCA (α-cyano-4-hydroxycinnamic acid) onto a MALDI plate (GOLD W/PIN, Applied Biosystems) for MS assay.

9. Use MALDI-MS (positive ion detection mode) to obtain the peak ratio (as determined by ratios of peak areas) of the ^{16}O- and ^{18}O-NAD^+ peaks (664/666) in the sample. Assay standards containing only ^{16}O and ^{18}O NAD^+ (600 pmol each) to determine corrections for isotopic purity and to provide calibration of the procedure.

10. Determine NAD^+ concentrations in samples by multiplying the corrected isotopic ratio 664/666 by the pmol of ^{18}O NAD added to the sample, dividing by cell number to generate pmol $NAD^+/10^6$ cells. Alternatively for mitochondria express as pmol NAD^+/mg protein.

4   Notes

1. For animal tissue samples, tissues need to be finely ground in liquid nitrogen before any treatment.

2. NADH can also be determined by cycling assay. For NADH measurement, samples should be mixed in 50 mM NaOH/1 mM EDTA, vortex and sonicate for three times, incubate at 60 °C for 30 min, cool on ice for 5 min, and then spin down, neutralize with 1 M HCl and 500 mM K_2HPO_4 (pH = 5.5) to pH = 7, determined by pH paper. When doing the cycling assay, use NADH instead of NAD^+ to prepare the standard curve (See detailed flow diagram in Fig. 4).

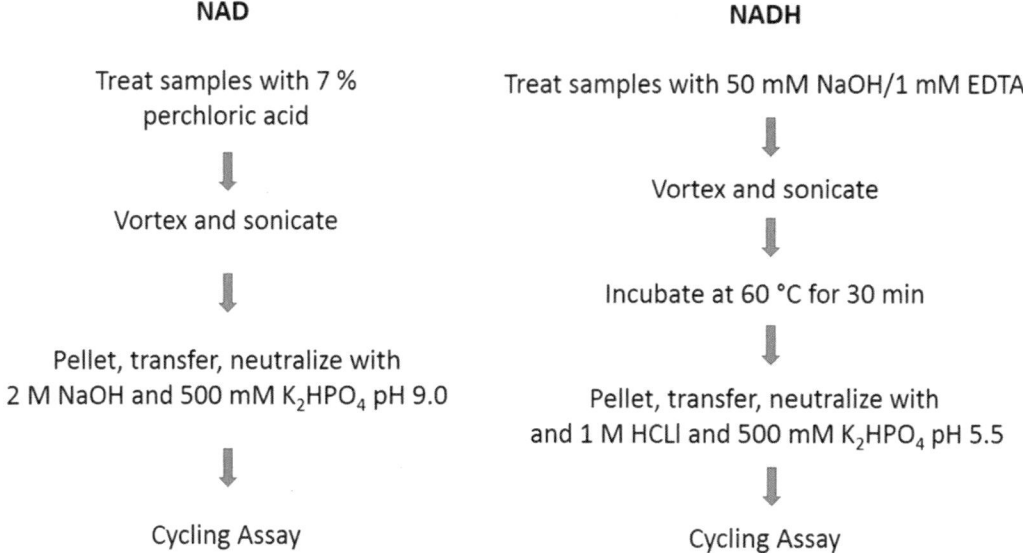

NAD

Treat samples with 7 % perchloric acid

⬇

Vortex and sonicate

⬇

Pellet, transfer, neutralize with 2 M NaOH and 500 mM K$_2$HPO$_4$ pH 9.0

⬇

Cycling Assay

NADH

Treat samples with 50 mM NaOH/1 mM EDTA

⬇

Vortex and sonicate

⬇

Incubate at 60 °C for 30 min

⬇

Pellet, transfer, neutralize with and 1 M HCLl and 500 mM K$_2$HPO$_4$ pH 5.5

⬇

Cycling Assay

Fig. 4 Comparison of NAD⁺ and NADH determination by cycling assay

3. As cycling assay is a time-sensitive assay, diaphorase ideally should be added to each well at the same time to initiate reaction. Make sure to use multichannel pipette at this step to reduce time error.

4. All the sample values should be in the range of the standard curve. If it is out of range, adjust the sample amount by diluting or adding more samples.

Acknowledgements

This work was supported by R01 DK74366, R21 DK094001-01A1, and 1R01 GM106072-01.

References

1. Sauve AA (2008) NAD+ and vitamin B3: from metabolism to therapies. J Pharmacol Exp Ther 324:883–893

2. Pollak N, Dolle C, Ziegler M (2007) The power to reduce: pyridine nucleotides–small molecules with a multitude of functions. Biochem J 402:205–218

3. Williamson DH, Lund P, Krebs HA (1967) The redox state of free nicotinamide-adenine dinucleotide in the cytoplasm and mitochondria of rat liver. Biochem J 103:514–527

4. Bai P, Canto C, Brunyanszki A, Huber A, Szanto M, Cen Y, Yamamoto H, Houten SM, Kiss B, Oudart H, Gergely P, Menissier-de Murcia J, Schreiber V, Sauve AA, Auwerx J (2011) PARP-2 regulates SIRT1 expression and whole-body energy expenditure. Cell Metab 13:450–460

5. Bai P, Canto C, Oudart H, Brunyanszki A, Cen Y, Thomas C, Yamamoto H, Huber A, Kiss B, Houtkooper RH, Schoonjans K, Schreiber V, Sauve AA, Menissier-de Murcia J,

Auwerx J (2011) PARP-1 inhibition increases mitochondrial metabolism through SIRT1 activation. Cell Metab 13:461–468

6. Canto C, Sauve AA, Bai P (2013) Crosstalk between poly(ADP-ribose) polymerase and sirtuin enzymes. Mol Aspects Med 34:1168–1201

7. Yang H, Yang T, Baur JA, Perez E, Matsui T, Carmona JJ, Lamming DW, Souza-Pinto NC, Bohr VA, Rosenzweig A, de Cabo R, Sauve AA, Sinclair DA (2007) Nutrient-sensitive mitochondrial NAD+ levels dictate cell survival. Cell 130:1095–1107

8. Haigis MC, Sinclair DA (2010) Mammalian sirtuins: biological insights and disease relevance. Annu Rev Pathol 5:253–295

9. Di Lisa F, Menabo R, Canton M, Barile M, Bernardi P (2001) Opening of the mitochondrial permeability transition pore causes depletion of mitochondrial and cytosolic NAD+ and is a causative event in the death of myocytes in postischemic reperfusion of the heart. J Biol Chem 276:2571–2575

10. Jacobson EL, Jacobson MK (1976) Pyridine nucleotide levels as a function of growth in normal and transformed 3T3 cells. Arch Biochem Biophys 175:627–634

11. Bembenek ME, Kuhn E, Mallender WD, Pullen L, Li P, Parsons T (2005) A fluorescence-based coupling reaction for monitoring the activity of recombinant human NAD synthetase. Assay Drug Dev Technol 3:533–541

12. Canto C, Houtkooper RH, Pirinen E, Youn DY, Oosterveer MH, Cen Y, Fernandez-Marcos PJ, Yamamoto H, Andreux PA, Cettour-Rose P, Gademann K, Rinsch C, Schoonjans K, Sauve AA, Auwerx J (2012) The NAD(+) precursor nicotinamide riboside enhances oxidative metabolism and protects against high-fat diet-induced obesity. Cell Metab 15:838–847

13. Yang T, Sauve AA (2006) NAD metabolism and sirtuins: metabolic regulation of protein deacetylation in stress and toxicity. AAPS J 8:E632–E643

Chapter 5

Measuring PGC-1α and Its Acetylation Status in Mouse Primary Myotubes

Ana P. Gomes and David A. Sinclair

Abstract

Metabolic flexibility is vital for the cells to adapt to different energetic situations, allowing the organisms to adapt to changing conditions and survive challenges. One of the most important regulators of the metabolic flexibility is PGC-1α activity. PGC-1α integrates numerous signals and regulates a variety of transcription factors and nuclear receptors that together regulate mitochondrial homeostasis and fatty acid oxidation. One of the major ways that PGC-1α activity is regulated is by changes in its acetylation status. Thus measuring the acetylation status of PGC-1α is an important indicator of the metabolic flexibility of the cells. In this chapter, we describe an approach to evaluate PGC-1α acetylation in primary mouse myotubes. The method is applicable to other cell types and tissues as well.

Key words PGC-1α, Acetylation, Sirtuin deacetylase, SIRT1, GCN5, Mitochondrial biogenesis, Respiration, Aging, Diabetes, Sarcopenia, Cachexia

1 Introduction

Peripheral tissues exhibit remarkable metabolic flexibility, a feature that is critical for maintenance of cellular homeostasis during changes in energy supply and demand. When energy levels are limited, as during fasting and exercise, these tissues have the ability to switch to use fatty acids as fuel for mitochondrial oxidation, preserving glucose for cells that strictly rely on glucose as their energy source [1]. Highlighting the importance of this flexibility, disturbances in metabolic homeostasis are frequently associated with several metabolic and age-related diseases [2–4].

Metabolic homeostasis is regulated by the coordinated activity of multiple transcription factors and nuclear receptors that tightly coordinate the expression of genes involved in mitochondrial metabolism and biogenesis [5, 6]. The peroxisome proliferator activated receptor gamma co-activator 1 alpha (PGC-1α) is one of the most central, integrating the activity of transcription factors that adjusts energy production, energy utilization, heat production, and

Carlos M. Palmeira and Anabela P. Rolo (eds.), *Mitochondrial Regulation*, Methods in Molecular Biology, vol. 1241, DOI 10.1007/978-1-4939-1875-1_5, © Springer Science+Business Media New York 2015

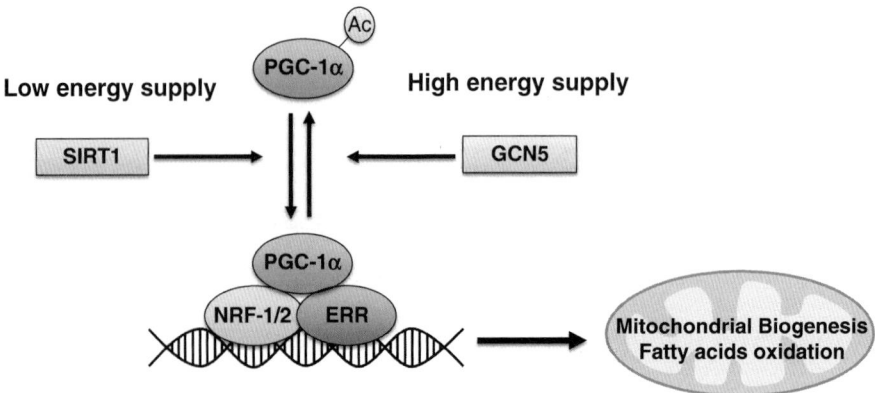

Fig. 1 The acetylation status of PGC-1α is a key regulatory switch in energy utilization. During low energy intake, such as fasting and calorie restriction, when there is a higher demand for fatty acid oxidation (FAO) and respiration, NAD+ levels increase, stimulating the activity of SIRT1 and decreasing the abundance of acetylated PGC-1α. Upon refeeding, the acetyltransferase GCN5, acetylates PGC-1α. The acetylation status of PGC-1α, the activity of SIRT1, and for evaluating potential new molecules for the treatment of mitochondrial diseases and metabolic disorders

tightly links the activities of genes in the nucleus with those in mitochondria. PGC-1α is also considered a "master regulator" of the mitochondrial biogenesis program [7, 8] that remodels muscle the fiber-type so that it is metabolically more oxidative and less glycolytic. Loss of PGC-1α signaling is considered a potential underlying cause of disorders such as diabetes, and cardiomyopathy, and obesity.

Regulation of PGC-1α activity is extremely complex and regulated at many levels by different signaling events [9]. One of the key regulatory events is PGC-1α acetylation status. Indeed, PGC-1α acetylation has been detected on 13 of its lysine residues, and acetylation negatively correlates with its activity. PGC-1α acetylation status is tightly co-regulated by the opposing activities of the acetyltransferase GCN5 [10, 11] and the deacetylase SIRT1 [12, 13]. GCN5 and SIRT1 are two important energetic sensors, as their activity is dependent on acetyl-coA and NAD+, respectively [14]. By monitoring the abundance of these cofactors, and controlling PGC-1α activity, they tightly regulate the ability of the cells to adapt their metabolism to the energy demands promoting metabolic flexibility (Fig. 1).

Acetylation of PGC-1α is an important marker of the energetic state of the cell, of mitochondrial biogenesis, and the relative activity of GCN5 and SIRT1. As such, the ability of measure acetylation is an important tool, not only in the study of upstream enzyme activity, but also for diseases related to metabolic inflexibility and evaluating the efficacy and mechanism of compounds aimed at treating metabolic diseases [15–18] Here we described a protocol to determine PGC-1α acetylation in primary myotubes, a key metabolic cell type that depends on PGC-1α for the relatively high abundance of mitochondria and rate of oxidative phosphorylation. We will describe methods to isolate primary myoblasts from mice and how to infect them with an adenovirus encoding Flag-PGC-1α that can be used to efficiently measure acetylation status.

2   Materials

All reagents should be the highest purity available and solutions should be prepared using distilled or ultrapure water.

2.1   Solutions and Media Formulations

1. Phosphate-buffered saline (PBS): 1 mM KH_2PO_4, 155 mM NaCl, 3 mM Na_2HPO_4-$7H_2O$, pH 7.4.

2. Collagenase/Dispase solution: 0.75 U/mL collagenase, 1.0 U/mL dispase, and 2.5 mM $CaCl_2$ diluted in PBS.

3. Pre-plating medium: 90 % low glucose Dulbecco's modified eagle medium (DMEM), 10 % FBS, Glutamax, antibiotic–antimycotic.

4. Growth medium: 80 % Ham's F-10 medium, 20 % FBS, Glutamax, 2.5 ng/mL bFGF, 10 ng/mL recombinant EGF, 1 µg/mL recombinant insulin solution, 0.5 mg/mL fetuin, 0.4 µg/mL dexamethasone, antibiotic–antimycotic (*see* **Note 1**).

5. Differentiation medium: 98 % low glucose DMEM, 2 % horse serum, Glutamax, 1 µg/mL recombinant insulin solution, antibiotic–antimycotic.

6. Immunoprecipitation buffer (IP buffer): 0.05 % NP-40, 50 mM NaCl, 0.5 mM EDTA, 50 mM Tris–HCl (pH 7.4), 10 mM nicotinamide, 1 µM trichostatin A, 2,000 U/mL Micrococcal nuclease, protease inhibitor cocktail.

7. Tris-buffered saline with Tween-20 10× (TBS-T): 0.2 M Tris base, 1.5 M NaCl, 0.05 % Tween 20, pH 7.4.

8. Blocking solution: 1 % BSA dissolved in 1× TBS-T buffer.

2.2   Other Equipment/Supplies

1. Flag-PGC-1α adenovirus and control adenovirus [12, 13].

2. Anti-PGC-1α antibody (Santa Cruz, H300).

3. Anti-FLAG M2-agarose affinity gel (Sigma-Aldrich).

4. Anti-Flag antibody (Sigma-Aldrich).

5. Anti-Pan acetylated lysine (Cell Signaling).

6. Clean-Blot IP Detection Reagent (HRP) (Pierce).

7. Flag peptide.

8. 2× Laemmli sample buffer.

9. Restore Western Blot Stripping buffer (Pierce).

10. Collagen coated cell culture plates.

11. Cell Scrappers.

12. Cell strainer 40 μm.

13. Equipment suitable to perform Western blot.

14. 4–20 % gradient Tris–HCl polyacrylamide gels.

15. Bovine serum albumin.

16. PVDF membrane.

17. ECL detection system.

18. Shaking water bath.

19. Orbital shaker.

20. Tube rotator.

21. Heat block with capacity to achieve 100 °C.

22. 37 °C incubator appropriated for tissue culture.

3   Methods

Experiments should be carried out at room temperature unless specified otherwise.

3.1   Isolation of Mouse Primary Satellite Cells

1. Sacrifice the animal and rapidly dissect out the hind limb muscles. Place the muscles in tissue culture dish containing PBS (*see* **Note 2**).

2. Transfer tissue to a new cell culture dish containing a small amount of PBS and mince tissue. Transfer the minced tissue to 50 mL tube containing 5 mL of PBS, and allow the tissue to settle.

3. Carefully discard the PBS on top of the minced tissue and add 2 mL of Collagenase/Dispase solution to the minced tissue (final volume for shaking ~5 mL).

4. Incubate for 2 h at 37 °C in a shaking water bath (speed should be set so visible agitation occurs) (*see* **Note 3**).

5. Add 10 mL of pre-plating media and gently triturate the tissue. Pass the cell suspension trough a cell strainer fit into a 50 mL conical tube.

6. Centrifuge the filtrate at $400 \times g$ for 5 min. Aspirate the supernatant and suspend the pellet in 7 mL of pre-plating media.

7. Transfer the tissue solution to a cell culture dish (non-collagen coated) and pre-plate for 30 min in a 37 °C incubator appropriated for tissue culture.

8. Remove non-adherent cells by gently swirling the dish and aspirating cell suspension into a 15 mL conical tube (*see* **Note 4**). Pellet the satellite cell by centrifuging the cell suspension at $300 \times g$ for 5 min.

9. Aspirate the supernatant and suspend the pellet in 15 mL of growth media.

10. Plate the satellite cells on collagen coated T-25 cm flasks for expansion in growth media and maintain them in a 37 °C incubator appropriated for tissue culture. Expand the cells until they are 80 % confluent and change 50 % of the media everyday.

11. After expansion the satellite cells can be differentiated into myotubes, to do so the cells should be plated and allowed to reach 80 % confluence. After this confluence is achieved switch the media to differentiation media and wait for 6 days replacing 50 % of the media daily (*see* **Note 5**).

3.2 Infection with Adenovirus Encoding Flag-PGC-1α and Empty Vector

1. Differentiate satellite cells into myotubes in 15 cm tissue culture dishes.

2. After differentiation of the satellite cells into myotubes, add the adenovirus to the media of the cells in a multiplicity of infection (MOI) of 2 and let the infection occur for 12 h.

3. Replace the differentiation media and replace with fresh differentiation media, wait for additional 36 h.

3.3 Immunoprecipitation and Western Blot

1. Remove the media from the cells and wash with 5 mL of PBS. Collect the samples by scrapping the cells in 2 mL of ice-cold IP buffer.

2. To lyse the cells snap freeze and thaw the cells three times.

3. Centrifuge the cell lysates at $12,000 \times g$ for 15 min at 4 °C.

4. Collect the supernatant to a new tube and quantify the proteins in the lysate by the Bradford method [19].

5. Add 500 μg of protein lysate and IP buffer to make a final volume of 500 μL. Keep an aliquot of the protein lysate to be used as input.

6. Prepare the Anti-FLAG M2-agarose affinity gel by pipetting 15 μL for each sample of the agarose gel into 500 μL of IP buffer and mix. Centrifuge at $500 \times g$ for 5 min. Remove the supernatant and add 100 μL of IP buffer per each sample that will be used.

7. Add 100 μL of the IP buffer containing the Anti-FLAG M2-agarose affinity gel prepared before to the protein lysate prepared on **step 5** and incubate overnight at 4 °C with agitation.

8. To wash the samples centrifuge the samples at $500 \times g$ for 5 min. Discard the supernatant, add 500 μL of IP buffer to the pellet and incubate for 15 min at 4 °C with agitation.

9. Repeat **step 8** for four additional times (*see* **Note 6**).

10. After washing the samples, elute the proteins by incubating the protein lysate in 50 μL of IP buffer containing 100 μg/mL of Flag peptide (*see* **Note 7**) and incubate for 1 h at 4 °C with agitation.

11. Centrifuge the samples at $12,000 \times g$ for 5 min and collect the supernatant.

12. Mix the supernatant and 1 % of the amount of the protein lysate used to do the initial IP (input) with 2X Laemmli sample buffer supplemented with 10 mM dithiothreitol (DTT) and boil for 5 min.

13. Load the samples and the input in 4–20 % gradient Tris–HCl polyacrylamide gels and run on SDS-PAGE under reducing conditions.

14. Transfer the separated proteins to a polyvinylidene difluoride membrane.

15. After the transfer is completed, incubate the membrane in 5 mL of blocking solution for 1 h with agitation.

16. Incubate overnight at 4 °C with agitation the membrane containing the IP samples in 5 mL of blocking solution containing the anti-pan-acetylated lysine antibody (dilution of 1:1,000) and the membrane containing the Input samples in 5 mL of blocking solution containing the anti-Flag antibody (dilution 1:5,000) (*see* **Note 8**).

17. Remove the blocking solution containing the primary antibodies and wash the membranes with 1× TBS-T for 20 min with agitation. Repeat this step three times.

18. Dilute Clean-Blot IP detection reagent with blocking solution and incubate for 1 h with agitation.

19. Remove the Clean-Blot IP detection reagent solution and wash the membranes with 1× TBS-T for 20 min with agitation. Repeat this step three times.

20. Prepare the ECL working solution by combining equal portions of the ECL Detection Reagent 1 and Reagent 2. Use 0.125 mL of working solution per cm^2 of membrane.

21. Add the ECL working solution to the blot and incubate for 5 min. Remove the ECL working solution and expose to film or CCD camera.

22. Wash the membranes in 1× TBS-T for 20 min with agitation.

23. Remove 1× TBS-T and add 5 mL of stripping Restore Western Blot stripping buffer. Incubate for 15 min with agitation.

24. Wash with 1× TBS-T until all the stripping solution is removed.

25. Incubate overnight at 4 °C with agitation the membrane containing the IP samples in 5 mL of blocking solution containing

the anti-Flag antibody (dilution of 1:5,000) and the membrane containing the Input samples in 5 mL of blocking solution containing the anti-PGC-1α antibody (dilution 1:500) (*see* **Note 8**).

26. Remove the blocking solution containing the primary antibodies and wash the membranes with 1× TBS-T for 20 min with agitation. Repeat this step three times.

27. Dilute Clean-Blot IP detection reagent with blocking solution and incubate for 1 h with agitation.

28. Remove the Clean-Blot IP detection reagent solution and wash the membranes with 1× TBS-T for 20 min with agitation. Repeat this step three times.

29. Prepare the ECL working solution by combining equal portions of the ECL Detection Reagent 1 and Reagent 2. Use 0.125 mL of working solution per cm^2 of membrane.

30. Add the ECL working solution to the blot and incubate for 5 min. Remove the ECL working solution and expose to film or CCD camera.

4   Notes

1. Growth factors and insulin should be added fresh to the media. Also avoiding freeze/thaw cycles of the aliquots is recommended.

2. To ensure a good yield of satellite cells, skeletal muscle of more than one mouse can be pulled together.

3. For maximum digestion, gently triturate the tissue one or two times during incubation with collagenase/dispase solution.

4. Fibroblasts preferentially adhere to plastic dishes, thus discarding this step can significantly affect cell purity. In addition, if purity is an issue in the cultures, repeating this step will help to make the cell population more pure.

5. Satellite cells will fuse during differentiation and long multinucleated myotubes will be visible. Time of differentiation might vary from culture to culture.

6. If high background and/or unspecific binding is observes, increasing the washing time will help reducing it.

7. The Flag-peptide is useful for competitive elution from the anti-flag antibody bound to the agarose gel. Alternatively, the samples can be boiled in 50 μL of 1× Laemmli Sample Buffer supplemented with 10 mM DTT for 5 min.

8. Optimization of the concentration of the antibodies might be required depending on the treatment of the cells and the purpose of the experiment.

Acknowledgements

We would like to thank Eric L. Bell for helpful advice on the immu-noprecipitation and acetylation detection protocol. This work was supported by a fellowship from the Portuguese Foundation for Science and Technology (SFRH/BD/44674/2008) to A.P.G, the Paul F. Glenn Foundation for Medical Research, the United Mitochondrial Disease Foundation, The Juvenile Diabetes Research Foundation, and NIA/NIH grants to D.A.S.

References

1. Storlien L, Oakes DE, Kelley DE (2004) Metabolic flexibility. Proc Nutr Soc 63(2):363–368. doi:10.1079/PNS2004349

2. Galgani JE, Moro C, Ravussin E (2008) Metabolic flexibility and insulin resistance. Am J Physiol Endocrinol Metab 295(5):E1009–E1017. doi:10.1152/ajpendo.90558.2008

3. Civitarese AE, Ravussin E (2008) Mitochondrial energetics and insulin resistance. Endocrinology 149(3):950–954. doi:10.1210/en.2007-1444

4. van den Brom CE, Huisman MC, Vlasblom R, Boontje NM, Duijst S, Lubberink M, Molthoff CF, Lammertsma AA, van der Velden J, Boer C, Ouwens DM, Diamant M (2009) Altered myocardial substrate metabolism is associated with myocardial dysfunction in early diabetic cardiomyopathy in rats: studies using positron emission tomography. Cardiovasc Diabetol 8:39. doi:10.1186/1475-2840-8-39

5. Scarpulla RC (2002) Nuclear activators and coactivators in mammalian mitochondrial biogenesis. Biochim Biophys Acta 1576(1–2):1–14. doi:10.1152/physrev.00025.2007

6. Scarpulla RC (2002) Transcriptional activators and coactivators in the nuclear control of mitochondrial function in mammalian cells. Gene 286(1):81–89. doi:10.1016/S0378-1119(01)00809-5

7. Puigserver P, Wu Z, Park CW, Graves R, Wright M, Spiegelman BM (1998) A cold-inducible coactivator of nuclear receptors linked to adaptive thermogenesis. Cell 92(6):829–839. doi:10.1016/S0092-8674(00)81410-5

8. Wu Z, Puigserver P, Andersson U, Zhang C, Adelmant G, Mootha V, Troy A, Cinti S, Lowell B, Scarpulla RC, Spiegelman BM (1999) Mechanisms controlling mitochondrial biogenesis and respiration through the thermogenic coactivator PGC-1. Cell 98(1):115–124. doi:10.1016/S0092-8674(00)80611-X

9. Fernandez-Marcos PJ, Auwerx J (2011) Regulation of PGC-1alpha, a nodal regulator of mitochondrial biogenesis. Am J Clin Nutr 93(4):884S–890S. doi:10.3945/ajcn.110.001917

10. Lerin C, Rodgers JT, Kalume DE, Kim SH, Pandey A, Puigserver P (2006) GCN5 acetyltransferase complex controls glucose metabolism through transcriptional repression of PGC-1alpha. Cell Metab 3(6):429–438. doi:10.1016/j.cmet.2006.04.013

11. Kelly TJ, Lerin C, Haas W, Sp G, Puigserver P (2009) GCN5-mediated transcriptional control of the metabolic coactivator PGC-1beta through lysine acetylation. J Biol Chem 284(30):19945–19952. doi:10.1074/jbc.M109.015164

12. Rodgers JT, Lerin C, Haas W, Gygi SP, Spiegelman BM, Puigserver P (2005) Nutrient control of glucose homeostasis through a complex of PGC-1alpha and SIRT1. Nature 434(7029):113–118. doi:10.1038/nature03354

13. Gerhart-Hines Z, Rodgers JT, Bare O, Lerin C, Kim SH, Mostoslavsky R, Alt FW, Wu Z, Puigserver P (2007) Metabolic control of muscle mitochondrial function and fatty acid oxidation through SIRT1/PGC-1alpha. EMBO J 26(7):1913–1923

14. Jeninga EH, Schoojans K, Auwerx J (2010) Reversible acetylation of PGC-1: connecting energy sensors and effectors to guarantee metabolic flexibility. Oncogene 29(33):4617–4624. doi:10.1038/onc.2010.206

15. Lagouge M, Argmann C, Gerhart-Hines Z, Meziane H, Lerin C, Daussin F, Messadeq N, Milne J, Lambert P, Elliott P, Geny B, Laakso M, Puigserver P, Auwerx J (2006) Resveratrol improves mitochondrial function and protects against metabolic disease by activating SIRT1

and PGC-1alpha. Cell 127(6):1109–1122. doi:10.1016/j.cell.2006.11.013

16. Feige JN, Lagouge M, Canto C, Strehle A, Houten SM, Milne JC, Lambert PD, Mataki C, Elliot PJ, Auwerx J (2008) Specific SIRT1 activation mimics low energy levels and protects against diet-induced metabolic disorders by enhancing fat oxidation. Cell Metab 8(5):347–358. doi:10.1016/j.cmet. 2008.08.017

17. Baur JA, Pearson KJ, Price NL, Jamieson HA, Lerin C, Kalra A, Prabhu VV, Allard JS, Lopex-Lluch G, Lewis K, Pistell PJ, Poosala S, Becker KG, Boss O, Gwinn D, Wang M, Ramaswamy S, Fishbein KW, Psencer RG, Lakatta EG, Le Couteur D, Shaw RJ, Navas P, Puigserver P, Ingram DK, de Cabo R, Sinclair DA (2006) Resveratrol improves health and survival of mice on a high-calorie diet. Nature 444(7117):337–342. doi:10.34 10/f.1052795.504713

18. Minor RJ, Baur JA, Gomes AP, Ward TM, Csiszar A, Mercken EM, Abdelmohsen K, Shin YK, Canto C, Scheibye-Knudsen M, Krawczyk M, Irusta PM, Martin-Montalvo A, Hubbard BP, Zhang Y, Lehrmann E, White AA, Price NL, Swindell WR, Pearson KJ, Becker KG, Bohr VA, Gorospe M, Egan JM, Talan MI, Auwerx J, Westphal CH, Ellis JL, Ungvari Z, Vlasuk GP, Elliott PJ, Sinclair DA, de Cabo R (2011) SRT1720 improves survival and healthspan of obese mice. Sci Rep 1:70. doi:10.1038/srep00070

19. Bradford MM (1976) A rapid and sensitive method for the quantitation of microgram quantities of protein utilizing the principle of protein-dye binding. Anal Biochem 72:248–254. doi:10.1016/0003-2697(76)905273

Measurement of Mitochondrial Oxygen Consumption Rates in Mouse Primary Neurons and Astrocytes

Sofia M. Ribeiro, Alfredo Giménez-Cassina, and Nika N. Danial

Abstract

The introduction of microplate-based assays that measure extracellular fluxes in intact, living cells has revolutionized the field of cellular bioenergetics. Here, we describe a method for real time assessment of mitochondrial oxygen consumption rates in primary mouse cortical neurons and astrocytes. This method requires the Extracellular Flux Analyzer Instrument (XF24, Seahorse Biosciences), which uses fluorescent oxygen sensors in a microplate assay format.

   Key words Oxygen consumption rate, Neurons, Astrocytes, Bioenergetics, Extracellular flux

1   Introduction

Accumulating evidence suggests that nutrient sensing pathways and mitochondrial fuel metabolism influence cellular responses to physiologic and pathophysiologic stress. Beyond generating ATP, mitochondrial oxidation of carbon substrates influences the cellular redox status (NADH/NAD ratio) and the availability of intermediary metabolites for anabolic processes [1]. Mitochondria are also pivotal in trafficking and buffering of intracellular calcium [2]. Neurons and glia are especially reliant on mitochondrial energy metabolism due to a high energy demand associated with the maintenance of ion homeostasis and calcium buffering required for proper regulation of neuronal electrical activity. Furthermore, several metabolic intermediates that are generated through mitochondrial metabolism serve as precursors for the synthesis of neurotransmitters [3].

   Although glucose is the preferred fuel for energy production in the brain, neural cells are also capable of metabolizing alternative fuels, including ketone bodies, lactate, and glutamine [4, 5]. Mitochondrial fuel oxidation leads to the generation of ATP through oxidative phosphorylation with concomitant consumption of oxygen. The evaluation of mitochondrial oxygen consumption

Carlos M. Palmeira and Anabela P. Rolo (eds.), *Mitochondrial Regulation*, Methods in Molecular Biology, vol. 1241, DOI 10.1007/978-1-4939-1875-1_6, © Springer Science+Business Media New York 2015

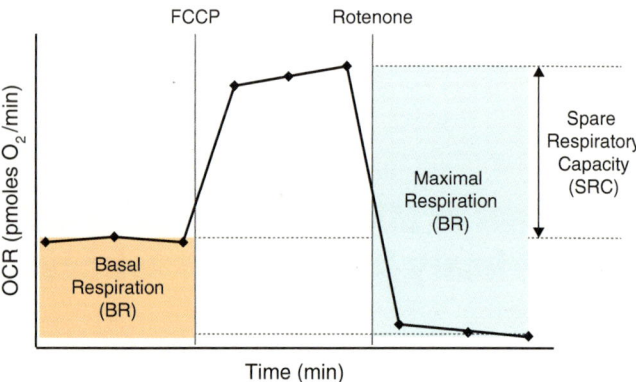

Fig. 1 Derivation of basal (BR) and maximal (MR) mitochondrial respiration rates as well as spare respiratory capacity (SRC). BR corresponds to the OCR in response to a given substrate that is responsible for ATP production and maintenance of the proton gradient. MR is the maximum achievable OCR that is induced upon treatment with a mitochondrial uncoupler, such as FCCP, and can indicate the maximal mitochondrial capacity to cope with increased metabolic demand. SRC is the extra reserve in mitochondrial respiratory capacity to respond to a raise in energy demand

rate (OCR) constitutes a direct measure of electron transport chain (ETC) activity, thus reflecting the efficiency with which mitochondria metabolize different fuels.

The assessment of mitochondrial function has particular relevance in neural cell types, given that many neurodegenerative diseases are associated with, or suggested to be triggered by, mitochondrial dysfunction [6–9]. The mitochondrial OCR and the bioenergetic parameters that can be derived from it represent an important measure of mitochondrial fitness. In a single OCR experiment, it is possible to determine important bioenergetic parameters, including basal respiration, maximal respiration, and spare respiratory capacity [6] (Fig. 1).

Mitochondrial OCR can be measured using a myriad of techniques that range from the classical Clark-type electrodes to fluorescent-based methods [10–12]. Classical methods utilizing the Clark-type electrode require a substantial amount of purified mitochondria, are relatively time consuming and do not allow automated injection of compounds [12]. The XF24 Extracellular Flux Analyzer (Seahorse Bioscience, Billerica, MA) uses a microplate-based assay to measure real-time OCR in intact, living cells, while it allows automated injection of useful compounds to determine bioenergetics parameters. It also allows the simultaneous comparison of multiple samples, increasing the reproducibility and biological significance of the assay [10–12].

In this chapter we describe detailed protocols that we have developed and applied to measure mitochondrial oxygen consumption rate in mouse cortical primary neurons and astrocytes using the XF24 extracellular flux analyzer [5]. The experimental paradigm described here assesses mitochondrial function in neural cells through the evaluation of important biogenetic parameters, such as basal respiration, maximal respiration, and spare respiratory capacity. This is especially useful to assess the mitochondrial respiration in response of different substrates.

2 Materials

Prepare all solutions in sterile conditions, using ultrapure water or the indicated solvents.

2.1 XF24 Analyzer Consumables

1. XF24 cell culture microplate.
2. XF24 extracellular flux assay kit. Each kit contains one sensor cartridge, one utility plate, and one lid.
3. Calibrant solution.

2.2 Culture Medium and Assay Medium

1. Astrocyte culture medium: 1:1 mix of DMEM (Life Technologies) + HAM F12 (Life Technologies), with 10 % FBS and 1 % Antibiotics (Pen/Strep).
2. Neuron culture medium: Neurobasal medium (Life Technologies) supplemented with 2 % B27, 2 mM Glutamax I and 1 % Antibiotics (Pen/Strep).
3. Assay medium: Unbuffered DMEM—For 1 L, mix one bottle of DMEM powder (8.3 g/L, Sigma-Aldrich) with 1.85 g of NaCl and bring it to ~900 mL with water. Adjust the pH to 7.4, and make up to 1 L with water. Add phenol red as a pH indicator (optional) (see **Note 1**).

2.3 Compounds for OCR Measurement

1. FCCP (Carbonyl cyanide 4-(trifluoromethoxy)phenylhydrazone) (Sigma-Aldrich): prepare stock solution at 0.5 mM in DMSO. Working concentration of FCCP can vary from batch to batch. We have found FCCP concentration of 1 µM for neurons and 0.5 µM for astrocytes to be appropriate for OCR studies [5] (see **Note 2**).
2. Rotenone: prepare stock solution at 5 mM in DMSO. Working concentration is 1 µM for both neurons and astrocytes.
3. Oligomycin: prepare stock solution at 5 mM in ethanol. Working concentrations usually vary between 0.5 and 1 µM (see **Note 3**).
4. The substrates to be used can be prepared in advance as stock aliquots. In Table 1, we indicate the solvents and concentrations we have previously used in our studies [5]. However, the working concentration of each substrate should be determined based on the specifics of each experiment (see **Note 1**).

Table 1
List of metabolic substrates and suggested working concentrations

Substrate	Solvent	Stock concentration	Working concentration[a]	Storage
Glucose	Water	1 M	10 mM	4 °C
β-Hydroxy-butyrate	Water	500 mM	5 mM	−20 °C
Lactate	1 M Hepes pH 7.4	1 M	5 mM	−20 °C
Glutamine[b]	Water	200 mM	5 mM	−20 °C
Pyruvate[c]	0.9 % NaCl	100 mM	1 mM	4 °C

[a]Suggested working concentration
[b]Commercially available (Life Technologies)
[c]Commercially available as a solution of sodium pyruvate (Sigma-Aldrich)

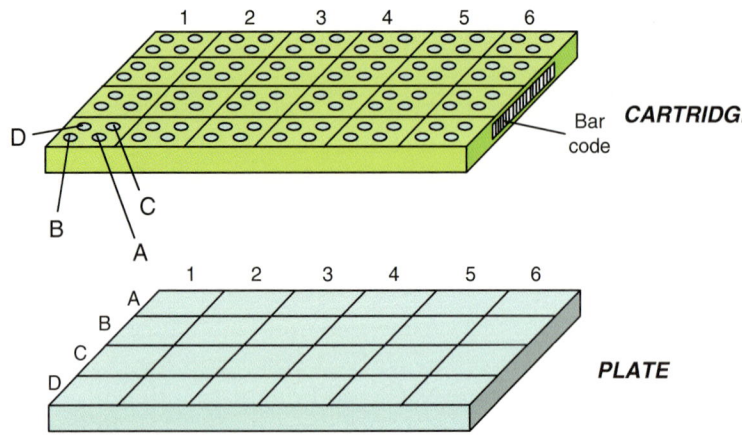

Fig. 2 Scheme of the XF24 microplate with the sensor cartridge

3   Methods

3.1   Seeding Primary Cortical Neurons and Astrocytes in XF24 Microplates

1. Leave at least two wells in the XF24 microplate unseeded, for background correction during OCR measurements; usually A1 and D6 (*see* Fig. 2).

2. For OCR measurements in primary neurons, obtain a suspension of neurons as previously described [5]. Seed 100 μL of a cell suspension in poly-L-lysine-coated XF24 microplates at a density of 1×10^5 cells/well, and let neurons differentiate for 5–6 days before the OCR experiment, replacing 1/3 of the culture medium with fresh medium, every 3 days (*see* **Notes 4** and **5**).

3. For OCR measurements in primary astrocytes, seed 100 μL of previously cultured astrocyte suspension in poly-L-lysine-coated XF24 microplates at a density of 4×10^4 cells per well [5]. Let cells recover overnight before the experiment (*see* **Notes 4** and **6**).

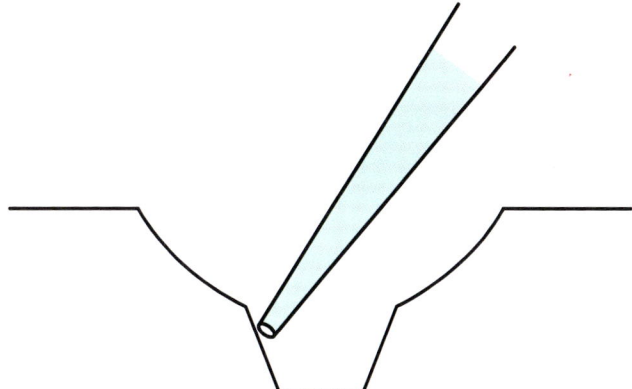

Fig. 3 Schematic representation of a single well from the XF24 microplate and the correct placement of the aspirating pipette during changes of medium, avoiding touching the layer of cells

4. The day before the experiment, hydrate cartridges (one cartridge per plate) with Seahorse XF24 calibrant solution. Incubate overnight in a CO_2-free incubator (*see* **Note 7**).

3.2 Assessment of Oxygen Consumption Rate in Neurons and Astrocytes Using the XF 24 Extracellular Flux Analyzer

1. On the day of the experiment, change the medium of the cells from culture medium to warm assay medium (DMEM without sodium bicarbonate, containing the desired fuel substrate) (*see* **Note 1**). For neurons, the assay medium may be supplemented with B-27 (*see* **Note 8**).

2. Start by carefully aspirating the medium from the plate using an aspirating pipette. Approach the plate from the side without touching the cells and remove the old medium leaving about 50–100 µL in each well (*see* Fig. 3) (*see* **Note 9**).

3. Add 300 µL of assay medium to each well to rinse the cells and carefully aspirate the medium as above.

4. Add 600 µL of assay medium to each well, including the background correction wells, and pre-incubate the cells for 1 h in a CO_2-free incubator.

5. While cells incubate, proceed to load the cartridge with the compounds to be injected. For the experimental paradigm described in this protocol, only two compounds are used, the mitochondrial uncoupler FCCP, and the complex I inhibitor rotenone. While loading the cartridge, make sure it is in the proper orientation (A–D are the rows and 1–6 are the columns, left to right, when the bar code is to the right side) (*see* Fig. 2).

6. Load 75 µL of FCCP in port A. Next, load 75 µL of rotenone in port B. If possible, leave at least one well per condition untreated, for baseline measurements. In these wells, load 75 µL of assay medium, similar to the background correction wells (*see* **Notes 10–12**).

7. The next step is to program the XF24 analyzer using the "assay wizard" function.

8. Follow the "assay wizard" instructions and fill out the information fields for each experiment. While most fields are not required to carry out the experiment, they are very important for the researcher's records and for proper data analysis. Therefore, it is strongly recommended to include all the information related to the experiment in each field.

9. Start by filling out the file name. The instrument's software is able to automatically add the date and time at which the cartridge was loaded to each file name. Select this option in the instrument's software.

10. Proceed through the next tabs filling out the suggested fields with the appropriate information for the experiment. In the "background correction" tab, define the background correction wells, which should correspond to the wells in the microplate where no cells were seeded (usually A1 and D6, *see* Fig. 2).

11. Proceed to the "groups and labels" tab. Define the experimental groups and attribute a group to each well. The background correction group should already be defined, but changes can be made at this point as required (*see* **Note 13**).

12. Define the protocol. The protocol described below is an example and can be modified to add more compounds and to adapt mix, measure and wait times to the cell preparation in use.

13. To measure OCR in primary neurons and astrocytes, using FCCP and rotenone as tool compounds to determine bioenergetic parameters, follow the protocol below:

    (a) Calibrate probes (*see* **Note 14**).

    (b) Equilibrate (*see* **Note 15**).

    (c) Mix—3 min (*see* **Note 16**).

    (d) Wait—2 min (*see* **Note 17**).

    (e) Measure—3 min (*see* **Note 18**).

    (f) Repeat **steps (c)–(e)**, three times, using the "loop" function

    (g) Inject port A, containing FCCP.

    (h) Repeat **steps (c)–(e)**, three times, using the "loop" function.

    (i) Inject port B, containing rotenone.

    (j) Repeat **steps (c)–(e)**, three times, using the "loop" function.

    (k) End protocol.

14. Once the instrument is programmed in the "assay wizard" tab, click on the "run" tab. Then, click "start." Before clicking on "start new plate," double check the name of the file and the folder directory where it will be saved. Any modifications to the fields noted above can be made at this point.

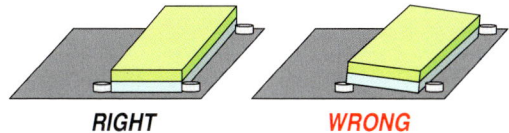

Fig. 4 Schematic representation of the proper placement of the XF24 calibration plate and cartridge in the XF24 analyzer tray

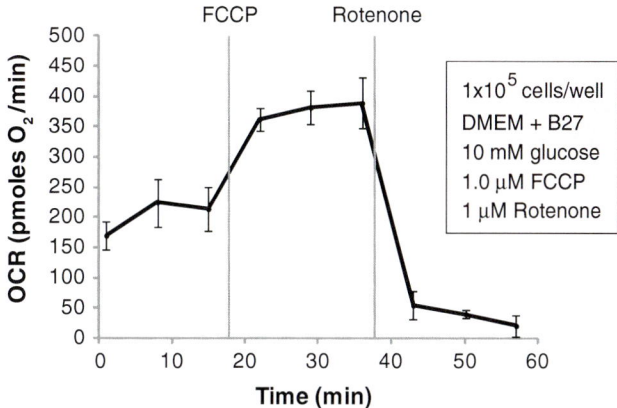

Fig. 5 Representative trace of real time oxygen consumption rates (OCR) in primary cortical neurons in response to glucose

15. Click on "start new plate." The lid will open and the tray will slide out. Place the cartridge with the utility plate on the tray, being careful to place it flat, and not on top of the holders. The bar code in the cartridge should face the back of the machine (*see* Figs. 2 and 4).

16. Press "continue." The tray should go back in and the lid will close. The calibration step takes approximately 25 min.

17. Once the calibration is finished, a message will appear on the screen, indicating that the microplate containing the cells can now be placed in the instrument. Press "OK," the lid will open and the tray will slide out with the utility plate, but not with the cartridge (which stays inside the instrument). Replace the calibrant plate for the microplate containing the cells and press "continue."

18. The instrument will now carry out the programmed protocol (*see* Figs. 5 and 6). Once the protocol ends, a message will appear on the screen to remove the plate with the cartridge. Press "OK," the lid will open and the tray will slide out. Remove the used microplate and the cartridge, and press "continuc" to closc thc lid.

19. Discard the cartridge. The cells in the microplate can be used for protein quantification and subsequent data normalization (*see* **Note 19**).

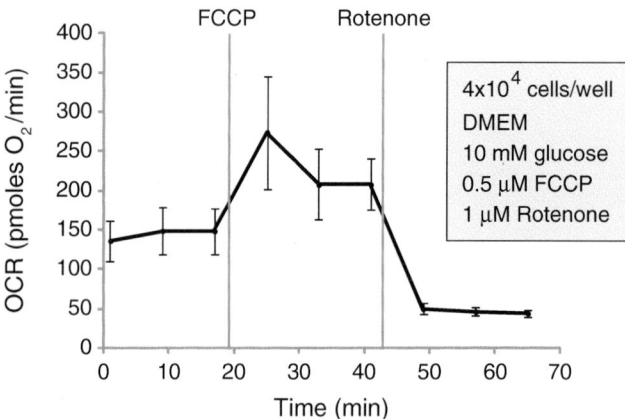

Fig. 6 Representative trace of real time oxygen consumption rates (OCR) in primary cortical astrocytes in response to glucose

3.3 Data Analysis   XF24 software uses an algorithm to calculate the oxygen consumption rates. There are two algorithms available: the original (Gain) Fixed algorithm and the new algorithm (Level) Direct (AKOS). With the release of the XF reader software version 1.7, the new "Direct" algorithm is the current default method to calculate oxygen consumption rates [11].

1. To analyze data and calculate the bioenergetic parameters, use the raw data that is retrieved in the excel data file. This file contains all the information from the experiment, including tables with OCR values for each individual well, in each time point.

2. To determine basal respiration (BR), calculate the difference between oxygen consumption rate in the beginning of the experiment, in basal conditions and the oxygen consumption rate at its lowest value, after addition of rotenone (*see* Fig. 1).

3. To determine maximal respiration (MR), calculate the difference between the oxygen consumption rate at its peak (after addition of FCCP), and the oxygen consumption rate at its lowest value, after addition of rotenone (*see* Fig. 1).

4. To determine the spare respiratory capacity (SRC), calculate the difference between the MR and the BR (*see* Fig. 1).

4   Notes

1. The substrate of choice can be added to the assay medium or be injected during the experiment using the injection ports. Having the substrate in the assay media will help define respiratory efficiency in the presence of that substrate. Injecting the substrate during the experiment will provide information on

the capacity of the cell to sense and respond acutely to a given substrate. For the latter, it is important to consider that not all cells have the ability to respond acutely to changes in substrate concentration and that cells should be kept at a very low basal respiratory rate to ensure the metabolic activity is not saturated and the acute response to the substrate is not masked. The choice of adding the substrate to the assay medium or injecting it during the experiment depends on the question being asked, the nature of the experiment, and the cellular context.

2. Each new aliquot of FCCP will have to be titrated for optimal working concentration.

3. Each new aliquot of Oligomycin will have to be titrated for optimal working concentration.

4. XF24 microplates should be pre-coated with poly-L-lysine (PLL) to ensure proper adherence of neurons and astrocytes to the microplate. For astrocytes, the culture dishes should also be pre-coated with PLL.

5. Seeding cells in a small volume allows the proper formation of a cell monolayer. In addition, neurons survive better in XF24 microplates when cultured in a small volume.

6. Primary astrocytes can be cultured for 2–3 weeks, but no more than two passages before the OCR experiment. This ensures culture homogeneity prior to the experiment.

7. Put enough volume of calibrant solution—500 µL to 1 mL—to completely immerse the sensors and incubate in CO_2 free incubator, at 37 °C. The optimal hydration time is ~16 h/overnight, but can go up to 72 h. If the hydration time exceeds 24 h, wrap the cartridge in parafilm to prevent evaporation of the calibrant solution.

8. Supplementing the assay medium with B-27 helps improve the viability of neurons and their response during the respirometry experiment.

9. When aspirating the medium, it is important to always leave some medium in each well to prevent cell detachment, air exposure, and potential cell death, thus ensuring an even monolayer of cells in all wells. It is important to be consistent with the volume left.

10. The recommended injection volume is 75 µL, but it can range from 50 to 100 µL.

11. The compounds to be injected should be pre-diluted in assay medium before loading into the sensor cartridge. It is necessary to pre-dilute the compounds to the proper concentration so that after injection, their concentration in the final volume of assay medium—after the addition of the injection volume—corresponds to the working concentration for each individual compound.

12. For each port that is selected for injection, the XF24 analyzer will inject the contents of that port in all wells. Even if the well is being used as a baseline well, it needs to contain running medium in the injection port, otherwise cell will get an air shot during the injection.

13. Labeling each well with the appropriate group information and color is important to ensure the correct grouping of the wells, both to follow the progression of the experiment in real time, and during data analysis using the XF24 software.

14. This step is always required to make sure the fluorescent sensors are working properly.

15. Once the plate has been transported from the incubator to the instrument, this step serves to equilibrate the cells to the temperature and pressure of O_2 and CO_2 in the instrument chamber.

16. Mixing time is required to replenish the oxygen content surrounding the cells. Every time the cartridge is lowered, a temporary chamber of 7 μL is formed on top of the cells. As cells consume oxygen, oxygen levels in the chamber drop and are subsequently measured by the fluorescent probes. After each measurement, the medium has to be mixed to replenish the oxygen content available to the cells and to ensure a homogeneous distribution of the compounds after each injection. During mixing, the instrument lowers and lifts the cartridge.

17. Waiting time is required to ensure the temperature is even and the cells are stable before subsequent measurements.

18. The shortest measuring time should be no less than 2 min. Measuring time should be adjusted for each cell type, considering their respiratory rates. It should allow sufficient time for the generation of a slope to calculate a consistent OCR value, corresponding to a total drop in the partial pressure of $O2 \leq 50$ mmHg.

19. To quantify protein from each individual well, add 10 μL of lysis buffer to each well after aspirating the assay medium and rinsing the cells with PBS. Scrape the cells with the tip of a P20 pipette and homogenize the lysate by pipetting up and down. Incubate the plate on ice for 15 min and then quantitate protein concentration using a plate-based assay, such as the BCA or the Bradford assays.

Acknowledgments

The authors acknowledge support from the US National Institute of Health grants R56 NS072142 and the Portuguese Foundation for Science and Technology (FCT) grant SFRH/BD/51200/2010. We thank the members of the Danial laboratory for helpful discussions.

References

1. Stanley IA, Ribeiro SM, Giménez-Cassina A et al (2013) Changing appetites: the adaptive advantages of fuel choice. Trends Cell Biol 24:118–127

2. Pivovarova NB, Andrews SB (2010) Calcium-dependent mitochondrial function and dysfunction in neurons. FEBS J 277(18):3622–3636

3. Mattson MP, Gleichmann M, Cheng A (2008) Mitochondria in neuroplasticity and neurological disorders. Neuron 60(5):748–766

4. Ronald ZH, Zielke CL, Baab PJ (2009) Direct measurement of oxidative metabolism in the living brain by microdialysis: a review. J Neurochem 109:24–29

5. Gimenez-Cassina A, Martinez-Francois JR, Fisher JK et al (2012) BAD-dependent regulation of fuel metabolism and K(ATP) channel activity confers resistance to epileptic seizures. Neuron 74(4):719–730

6. Brand MD, Nicholls DG (2011) Assessing mitochondrial dysfunction in cells. Biochem J 435(2):297–312

7. Lange M, Zeng Y, Knight A et al (2012) Comprehensive method for culturing embryonic dorsal root ganglion neurons for seahorse extracellular flux XF24 analysis. Front Neurol 3:175

8. Sauerbeck A, Pandya J, Singh I et al (2011) Analysis of regional brain mitochondrial bioenergetics and susceptibility to mitochondrial inhibition utilizing a microplate based system. J Neurosci Methods 198(1):36–43

9. Clerc P, Polster BM (2012) Investigation of mitochondrial dysfunction by sequential microplate-based respiration measurements from intact and permeabilized neurons. PLoS One 7(4):e34465

10. Wu M, Neilson A, Swift AL et al (2007) Multiparameter metabolic analysis reveals a close link between attenuated mitochondrial bioenergetic function and enhanced glycolysis dependency in human tumor cells. Am J Physiol Cell Physiol 292(1):C125–C136

11. Gerencser AA, Neilson A, Choi SW et al (2009) Quantitative microplate-based respirometry with correction for oxygen diffusion. Anal Chem 81(16):6868–6878

12. Ferrick DA, Neilson A, Beeson C (2008) Advances in measuring cellular bioenergetics using extracellular flux. Drug Discov Today 13(5–6):268–274

Chapter 7

Determination of Oxidative Phosphorylation Complexes Activities

João S. Teodoro, Carlos M. Palmeira, and Anabela P. Rolo

Abstract

Mitochondria possess a genome that codes for proteins, in the same fashion as the nuclear genome. However, the small, circular mitochondrial DNA (mtDNA) molecule has a reduced base pair content, for it can only code for 2 rRNA, 22 tRNA molecules, and 13 proteins, all of them part of the mitochondrial respiratory chain. As such, all of the other mitochondrial components derive from nuclear genome. This separation leads to a requirement for a well-tuned coordination between both genomes, in order to produce fully functional mitochondria. A vast number of pathologies have been demonstrated to involve, to some extent, alterations in mitochondrial function that, no doubt, can be caused by alterations to the respiratory chain activity. As such, several methods and techniques have been developed to assess both content and function of mitochondrial proteins, in order to help understand mitochondrial involvement on the pathogenesis of disease. In this chapter, we will address some of these methods, with the main focus being on isolated mitochondria.

Key words Mitochondria, Respiratory chain, Polarography, Mitochondrial proteic complexes, Spectrophotometry

1 Introduction

Mitochondria are the eukaryotic cell's main site of ATP generation, accounting for roughly 95 % of all ATP consumed in the cell. Along with ATP generation, mitochondria are also responsible for the storage of various molecules and ions (as, for example, Ca^{2+}). It is also here, in a physically separated (from the cytosol) environment that several biochemical reactions can take place, namely the lipidic β-oxidation, the citric acid cycle (or Krebs cycle), to name a few. As such, it comes as no surprise that mitochondrial data are extremely important to a variety of studies.

The oxidative phosphorylation system is made up from several proteic complexes (I to IV), which together with the mobile proteins ubiquinone (also known as coenzyme Q_{10}) and cytochrome c make the respiratory chain, across the mitochondrial inner membrane.

Carlos M. Palmeira and Anabela P. Rolo (eds.), *Mitochondrial Regulation*, Methods in Molecular Biology, vol. 1241, DOI 10.1007/978-1-4939-1875-1_7, © Springer Science+Business Media New York 2015

This system takes the electrons from metabolic substrates and transports them in increasingly free-energy favorable electronic jumps towards O_2, forming H_2O at Complex IV. Since this electronic transport is energetically favorable, a lot of energy is released. These proteic complexes (except Complex II) are assembled in such a way that this energy is harnessed to eject protons from the mitochondrial matrix, across the proton-impermeable inner membrane into the intermembrane space. This proton gradient can then be used by the fifth complex, the ATPSynthase to drive the phosphorylation of ADP into ATP.

The respiratory chain proteic complexes constituents are coded in both mitochondrial and nuclear genomes, and mutations on these genes are the most common reason for mitochondrial diseases, from diabetes to neurological diseases and aging [1]. Here, we describe one protocol for the characterization of the activity of all these proteic complexes in isolated mitochondria from tissue homogenates.

2   Materials

2.1   Reagents and Buffers

Standard hepatic mitochondrial homogenization buffer:

- Sucrose 250 mM.
- EGTA 0.5 mM.
- Bovine Serum Albumin 0.5 %.
- HEPES 10 mM, pH 7.4.

Standard hepatic mitochondrial wash buffer:

- Sucrose 250 mM.
- HEPES 10 mM, pH 7.4.

Standard hepatic mitochondrial respiratory buffer:

- Sucrose 130 mM.
- KCl 50 mM.
- $MgCl_2$ 5 mM.
- KH_2PO_4 5 mM.
- HEPES 10 mM, pH 7.4.

Standard ATPase reaction buffer:

- Sucrose 125 mM.
- KCl 65 mM.
- $MgCl_2$ 2.5 mM.
- HEPES 50 mM, pH 7.2.

Biuret reagent:

To make Biuret reagent, weight 1.5 g of cupric sulfate pentahydrate ($CuSO_4.5H_2O$) and dissolve it in 500 mL of water. Then, weight 6 g of sodium potassium tartrate tetrahydrate ($NaKC_4H_4O_6.4H_2O$) and slowly add it to the dissolved cupric sulfate solution (*see* **Note 1**). Add 300 mL of a 10 % solution of sodium hydroxide (NaOH). Add water to a final volume to 1 L, store in a light-protected glass bottle. It should be stable for a year.

Molybdate reagent:

To produce Molybdate reagent, dissolve 5 g of ferrous sulfate ($FeSO_4$) in 60 mL of water. After which, add 10 mL of a 10 % ammonium molybdate (($NH_4)_6Mo_7O_{24}.4H_2O$) solution. Adjust the volume to 100 mL with water.

Reagents:

- 2,6-dichloroindophenol (DCIP).
- 5,50′-dithiobis-2-nitrobenzoic acid (DTNB).
- Acetyl-CoA.
- Antimycin A.
- Ascorbate.
- ATP.
- Bovine Serum Albumin (BSA).
- Decylubiquinone.
- Deoxycholic acid (DOC).
- HCl.
- KCN.
- KH_2PO_4.
- $KHCO_3$.
- Lithium borohydrate.
- NADH.
- Oxaloacetate.
- Oxidized cytochrome *c*.
- Phenazine methosulfate.
- Rotenone.
- Succinate.
- Tetramethyl-phenylenediamine (TMPD).
- Trichloroacetic acid (TCA).
- Tris.
- Triton X-100.
- Ultra-pure water.

2.2 Equipment Required

- 50 mL centrifuge tubes.
- Animal (Rat, mice, other).
- Clark-type oxygen electrode, reaction chamber and associated register.
- Eppendorf-like 1.5 mL tubes.
- Glass Potter-Elvejhem homogenizer with Teflon pestle.
- Laboratory chronometer.
- Magnetic stirrer and magnetic stir bar.
- Orbital vortex.
- Plastic cuvettes.
- Power drill.
- Precision pipettes.
- Refrigerated centrifuge.
- Small volume beakers.
- Smooth paintbrushes.
- Spectrophotometer with cuvette access.
- Surgical scissors.
- Surgical tweezers.
- Test tube holder.
- Test tubes.
- Water bath, preferably with lid.

3   Methods

3.1 Isolation of Hepatic Mitochondria

- Take the animal, starved overnight, and quickly sacrifice it by cervical dislocation and decapitation. Bleed the animal into the sink for 5–10 s.
- Using surgical scissors and tweezers, cut open the animal's abdomen, right below the ribcage and remove the liver, in large pieces, into a beaker containing ice-cold homogenization buffer (*see* **Note 2**). Remove any adhering fat, fibrous tissue or blood vessels from the liver chunks, and thinly chop the liver into small pieces (*see* **Note 3**). Replace buffer.
- Add approximately 5–6 mL of ice-cold homogenization buffer to each gram of chopped liver (typically 50 mL) and transfer this mix into a precooled glass Potter-Elvejhem homogenizer.
- Homogenize the tissue using three to four up/down strokes of pestle rotating at roughly 300 rpm (*see* **Note 4**).
- When the tissue is homogenized, transfer the homogenate into two previously cooled centrifuge tubes, balancing them with

Table 1
Biuret Assay preparation

Tube	H$_2$O (µL)	BSA 0.4 % (µL)	Wash buffer (µL)	Sample (µL)	DOC 10 % (µL)	Biuret reagent (mL)	Protein content (mg)
0	500	0	50	–	50	2	0.0
1	250	250	50	–	50	2	1.0
2	125	375	50	–	50	2	1.5
3	0	500	50	–	50	2	2.0
X$_1$	500	–	–	50	50	2	X
X$_2$	500	–	–	50	50	2	X

BSA bovine serum albumin, *DOC* deoxycholic acid

homogenization buffer. Centrifuge the homogenate at $800 \times g$ for 10 min, at 4 °C (*see* **Note 5**).

- Carefully decant the supernatant into new cooled centrifugation tubes (*see* **Note 6**). Balance them with homogenization buffer and centrifuge at $10,000 \times g$ for 10 min at 4 °C.

- Discard the supernatant as completely as possible and gently resuspend the pellet in a small volume (roughly 3–5 mL) of wash buffer (*see* **Note 7**).

- Repeat the last step two more times. After the final centrifugation (third at $10,000 \times g$, fourth total) resuspend the mitochondrial pellet into a small volume of wash buffer (*see* **Note 8**).

- Let the mitochondria rest for 15–30 min (*see* **Note 9**).

3.2 Mitochondrial Quantification

- During the mitochondrial rest period described above, it is possible to use this time to quantify the mitochondrial protein content, since all assays are normalized against the mitochondrial protein content. For isolated rat mitochondria, we recommend the use of the Biuret assay [2]. Briefly, follow Table 1. Tubes 0–3 are for the standard curve.

- Prepare the tubes as in Table 1 (*see* **Note 10**). Mix the tubes in an orbital vortex and put them in a water bath, light protected at 37 °C for 5 min.

- When the 5 min are over, measure the absorbance of the tubes' content in standard plastic 4 mL cuvettes at 540 nm in a standard spectrophotometer (*see* **Note 11**). Construct a standard curve from the 1–3 tubes.

- Calculate the average absorbance of your duplicates (*see* **Note 12**) and insert it the following formula:

$$X = \left(\left(\text{Abs} - \text{Ins}\right) / \text{Slope}\right) \times \text{DF}$$

Where Abs is the average absorbance of your samples, Ins is the standard curve intersection, Slope is the standard curve slope and DF is the dilution factor. X is your sample's protein content, expressed in mg/mL (*see* **Note 13**).

3.3 Citrate Synthase Activity Assay

Citrate synthase is a mitochondrial matrix enzyme. This assay is used as an indicator if mitochondrial numbers in tissue homogenates and impure mitochondrial preparations, as it has not been found to be deficient in any disease states. Respiratory chain enzyme activities can be expressed as citrate synthase ratios to correct for any variations in mitochondrial numbers in these preparations. Also, mitochondrial integrity can assessed using fresh mitochondria by this assay by citrate synthase latency in the presence of Triton X-100.

This assay monitors the release of free coenzyme A (CoA) form Acetyl-CoA after the citrate synthase reactions initiated by the addition of oxaloacetate. This is achieved by Ellman's reagent (5,50′-dithiobis-2-nitrobenzoic acid, DTNB) reaction with the free thiol groups of CoA and registering the absorbance at 412 nm.

- Prepare the following reagents:
  1. Tris 200 mM, pH 8.0.
  2. Acetyl-CoA 100 mM.
  3. DNTB 100 mM (4 mg/mL)—add a pinch of $KHCO_3$ to help dissolve.
  4. Oxaloacetate 10 mM (1.4 mg/mL) add Tris to pH 7.0.
  5. Triton X-100 10 %.

- For each sample, the reaction mix is made in a cuvette from 500 μL Tris, 20 μL Acetyl-CoA, 20 μL DNTB, and 50 μg of mitochondrial protein. Add water to make 1 mL. After a light mix, record a stable baseline at 412 nm.

- Start the reaction by adding 10 μL of oxaloacetate (*see* **Note 14**).

- Finally, add 0.1 % (v/v—final concentration) Triton X-100 after recording the oxaloacetate rate (*see* **Note 15**).

With fresh mitochondria, the addition of oxaloacetate will cause an increase in the absorbance rate as intense as more citrate synthase is in the medium, meaning, the higher the rate, the more mitochondria are damaged. Adding Triton releases all of the citrate synthase still trapped inside mitochondria, which should demonstrate a steady value between preparations.

The extinction coefficient for DTNB at 412 nm is of 13.6 M/cm, which means that citrate synthase activity is calculated as follows:

$$\text{Activity}\left(\text{nmol.min.mL}\right) = \left(\Delta\text{Abs} \times \text{Vt} \times 1{,}000{,}000\right) / \left(\text{EC} \times \Delta\text{time} \times \text{mg} \times C_{[S]}\right)$$

where ΔAbs is the change in absorbance, Vt is the final cuvette volume in microliters, mg is the mitochondrial protein content (for this example, it is 0.05), EC is the extinction coefficient (in this example, 13.6), $C_{[S]}$ is the sample concentration, and Δtime is the elapsed time between stable recordings of absorbance.

3.4 Individual Respiratory Chain Enzymatic Activities

3.4.1 Complex I (NADH:Ubiquinone Oxidoreductase)

Complex I is the first proteic complex of the mitochondrial oxidative phosphorylation system. It receives electrons from NADH (which is oxidized to NAD^+) and transports them to the oxidized coenzyme-Q_{10}, or ubiquinone. To this oxidation of NADH and electron transport to ubiquinone is coupled the vectorial ejection of protons from the mitochondrial matrix to the intermembrane space, creating a protonic gradient with an electric and pH components, which is used by the ATPSynthase to drive the phosphorylation of ADP into ATP. It is the largest and heaviest complex of the respiratory chain, and the deficiency in Complex I is probably the most frequently encountered cause of mitochondrial disease, and various mutations in the genes coding for both the nuclear and mitochondrially encoded subunits have already been described [3]. Here, we describe a method first published by Janssen and collaborators [4]. As the authors describe, this method relies on Complex I oxidizing NADH and giving the electrons to the artificial substrate decylubiquinone, which in turn delivers them to 2,6-dichloroindophenol (DCIP). It is then possible to spectrophotometrically follow the reduction of DCIP by measuring the absorbance at 600 nm. It is a highly sensitive method, since decylubiquinone does not accept electrons from other sources [5]. This method also has the advantage of not requiring an UV-light source, for by following the disappearance rate of NADH as described by other methods [1], one would require UV-ready cuvettes and a spectrophotometer with an UV lamp, which is harder and much more expensive to obtain. As such, recording at 600 nm, within the visible spectra, one can bypass such issues.

- Prepare respiration buffer (as described in Subheading 2.1).
- To this solution, BSA 0.4 % should be added to maximize Complex I activity [4]. Also add KCN 240 μM, Antimycin A 4 μM, DCIP 60 μM, and decylubiquinone 70 μM. Prepare two tubes per sample, one with and other without rotenone 2 μM (*see* **Note 16**).
- Add 0.3 mg of freeze-thawed mitochondrial protein to both tubes. Record the absorbance at 600 nm, 25 °C, to register a steady rate, and then add to both tubes freshly prepared NADH 10 mM. Mix lightly and record the absorbance rate.

Since the extinction coefficient for DCIP is 19.1 M/cm, the specific activity of Complex I is given by:

$$\text{Activity}\left(\text{nmol.min.mL}\right) = \left(\left(\varDelta\text{Abs}_{[\text{S}]} - \varDelta\text{Abs}_{[\text{Blk}]}\right) \times \text{Vt} \times 1{,}000\right) / \left(\text{EC} \times \varDelta\text{time} \times \text{mg} \times C_{[\text{S}]}\right)$$

where $\varDelta\text{Abs}_{[\text{S}]}$ is the change in sample absorbance, $\varDelta\text{Abs}_{[\text{Blk}]}$ is the change in rotenone tube absorbance, Vt is the final cuvette volume in microliters, mg is the mitochondrial protein content (for this example, it is 0.03), EC is the extinction coefficient (in this example, 19.1), $C_{[\text{S}]}$ is the sample concentration, and $\varDelta\text{time}$ is the elapsed time between stable recordings of absorbance.

3.4.2 Complex II (Succinate:Ubiquinone Oxidoreductase)

Complex II is a particular proteic complex, which has four major differences from the other complexes of the respiratory chain. Its subunits are coded exclusively in the nuclear genome, is not a transmembrane protein (i.e., it is only present in the matrix side of the inner mitochondrial membrane and in its interior), its activity does not result in direct proton ejection (although the electrons it collects from oxidizing succinate can lead to protonic ejection at the level of Complexes III and IV) and, as it name implies, it is also part of the Krebs cycle, or Citric Acid Cycle. To evaluate the activity of Complex II, we describe a method first published by [6]. Contrary to Complex I (and III and V, as seen later), the activities of Complexes II and IV are polarographically evaluated, with resource to a Clark-type oxygen electrode, connected to a digital or analogic register. In this method, Complex II accepts electrons from Succinate and, since mitochondria are burst due to the freeze-thaw cycle, it does not donate electrons to ubiquinone, but rather to the supplied acceptor, phenazine methosulfate (PMS). PMS will then, in turn, donate the electrons directly to molecular oxygen, causing its conversion to water.

- Prepare respiration buffer as above (described in Subheading 2.1).
- Depending on the volume of the oxygen electrode chamber, use at least 1 mL of respiration buffer supplemented with:
  1. Succinate 5 mM.
  2. Rotenone 2 µM (Complex I inhibitor).
  3. Antimycin A 0.1 µg/mL of final volume (Complex III inhibitor).
  4. KCN 1 mM (Complex IV inhibitor).
  5. Triton X-100 0.3 mg/mL of final volume.
  6. 0.3 mg freeze-thawed mitochondrial protein.
- The reaction is initiated by the addition of PMS 1 mM (*see* **Note 17**).
- Complex II activity is measured by calculating the slope of oxygen decrease registered by the electrode, in its initial phase, as demonstrated by Fig. 1, and is typically presented in nAtoms O/min/mg protein (*see* **Note 18**).

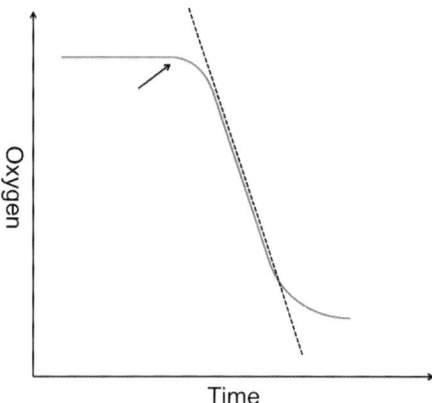

Fig. 1 Typical polarographic register for the evaluation of Complexes II and IV activities. The *red line* denotes the oxygen presence in the chamber. The *arrow marks* the addition of the reaction initiator (PMS for II and ascorbate + TMPD for IV). The *dashed line* is then drawn to calculate the slope of oxygen disappearance (Color figure online)

3.4.3 Complex III (Coenzyme Q: Cytochrome c Oxidoreductase)

Complex III of the mitochondrial respiratory chain is responsible for transferring electrons from fully reduced ubiquinone (known as ubiquinol) to the soluble protein cytochrome *c*, with a consequent protonic ejection. As mentioned before, the activity of Complex III can be spectrophotometrically evaluated using the method first described by [7], and further explored by [1]. In this method, similarly as before, one follows the increase in absorbance at 550 nm caused by the formation of reduced cytochrome *c*.

- Prepare respiration buffer as above (described in Subheading 2.1).

- Use a solution of decylubiquinone to freshly prepare decylubiquinol (*see* **Note 19**).

- To 1 mL of respiration buffer, add 0.3 mg of freeze-thawed mitochondrial protein, decylubiquinol 80 mM, KCN 240 μM, rotenone 4 μM, and ATP 200 μM. Prepare two tubes, one with and the other without Antimycin A 0.1 μg/mL of final volume (*see* **Note 20**).

- Start the reaction by adding oxidized cytochrome *c* 40 μM and then measure the rate of absorbance.

Since the extinction coefficient for reduced cytochrome *c* is 0.021 M/cm, the specific activity of Complex III is given by:

$$\text{Activity (nmol.min.mL)} = \left(\left(\Delta\text{Abs}_{[S]} - \Delta\text{Abs}_{[Blk]}\right) \times \text{Vt} \times 1{,}000\right) / \left(\text{EC} \times \Delta\text{time} \times \text{mg} \times C_{[S]}\right)$$

where $\Delta\text{Abs}_{[S]}$ is the change in sample absorbance, $\Delta\text{Abs}_{[Blk]}$ is the change in rotenone tube absorbance, Vt is the final cuvette volume in microliters, mg is the mitochondrial protein content (for this example, it is 0.03), EC is the extinction coefficient (in this example, 0.021), $C_{[S]}$ is the sample concentration, and Δtime is the elapsed time between stable recordings of absorbance.

3.4.4 Complex IV (Cytochrome c Reductase)

Complex IV is the last proteic complex of the electronic transport chain. It receives electrons from cytochrome c and supplies them to molecular oxygen, generating water. This electric transport, as with Complexes I and III, is accompanied by the protonic ejection from the matrix to the intermembrane space. As with Complex II, Complex IV's activity can be polarographically evaluated using the method first described by [8]. This method also implies oxygen consumption, but since that is a natural reaction for Complex IV, no artificial electron acceptor is required. The reaction is also supplemented with a mix of ascorbate + tetramethylphenylenediamine (TMPD) (*see* **Note 21**).

- Prepare respiration buffer as above (described in Subheading 2.1).
- Depending on the volume of the oxygen electrode chamber, use at least 1 mL of respiration buffer supplemented with:
  1. Rotenone 2 μM (Complex I inhibitor).
  2. Oxidized cytochrome c 10 μM.
  3. Triton X-100 0.3 mg/mL of final volume.
  4. 0.3 mg freeze-thawed mitochondrial protein.
- Start the reaction by adding Ascorbate 5 mM + TMPD 0.25 mM (*see* **Note 22**).
- As with Complex II, activity should be evaluated by calculating the slope of oxygen decrease registered by the electrode, in its initial phase, as demonstrated by Fig. 1, and is typically presented in nAtoms O/min/mg protein.

3.4.5 ATPase Activity of Mitochondrial ATPSynthase

Mitochondrial ATPSynthase in the enzymatic complex that utilizes the electrochemical gradient generated by the respiratory chain to drive ADP phosphorylation. The potential energy stored in such gradient is utilized by ATPSynthase to drive the generation of a chemical bond between ADP and a free phosphate group. The protons are then returned to the matrix, where they can be used, either by the respiratory chain or other enzymes inside mitochondria. It has been shown that, when the membrane potential is sufficiently reduced or completely destroyed, ATPSynthase can revert its activity, i.e., it has been shown that it will utilize ATP to pump protons out of the mitochondrial matrix [9]. This property can be explored to assess the activity of this enzymatic complex.

- Produce ATPase reaction buffer as above (described in Subheading 2.1).

Table 2
ATPase reaction tubes

Tube	Reaction buffer (mL)	Mitochondria (mg)	Oligomycin (μg)
X1	2	0.25	0
X1'	2	0.25	0.25

Table 3
ATPase reaction standard curve

Tube	KH$_2$PO$_4$ 0.5 mM (mL)	H$_2$O (mL)	nmol of Pi in tube
0	0	3	0
1	0.25	2.75	125
2	0.5	2.5	250
3	1	2	500

- Produce Trichloroacetic acid (TCA) 40 % (v/v), KH$_2$PO$_4$ 0.5 mM, and Molybdate reagent 10 % in H$_2$SO$_4$ 10 N (as described in Subheading 2.1).

- Prepare the following tubes, two for each sample (*see* **Note 23**) (Table 2).

- Place the tubes in a water bath at 37 °C. Start the reaction in each tube with ATP 2.5 mM.

- After precisely 10 min, stop each reaction with 1 mL TCA. Mix with an orbital vortex and put the tubes on ice. Collect 1 mL (of the three inside the tube) into a new tube.

- At this time, prepare a standard curve using Table 3.

- After preparing all the tubes (samples and standard curve), add 2 mL of Molybdate reagent to each. Vortex and let react for 3 min on the bench. Read absorbance at 660 nm.

The ATPase activity is calculated as before, for Biuret reaction. Plot the standard curve absorbance against Table 3 supplied Pi values. Again, an r^2 value under 0.995 indicates a non-acceptable curve. Calculate the slope and the intercept. Subtract the intercept to the sample's absorbance and divide the result by the slope. This value is the nmol of Pi inside the tube. Multiply this value by 4 (to normalize to 1 mg of protein) and then again by 3 (because of the 3 mL inside the first reaction tube only one was used) and divide the result by 10 (to normalize to 1 min). Do this to all sample tubes, and subtract the oligomycin tubes to their respective, no-oligomycin tube. The result is the specific activity of the mitochondrial ATPase. In sum,

$$\text{Mitochondrial ATPase activity} \left(\text{nmol Pi} / \text{mg protein} / \text{min}\right)$$
$$= \left(\left(\left(\text{Abs}_{[S]} - \text{Ins}\right) / \text{Slp}\right) \times 1.2\right) - \left(\left(\left(\text{Abs}_{[O]} - \text{Ins}\right) / \text{Slp}\right) \times 1.2\right)$$

where $\text{Abs}_{[S]}$ is the sample absorbance, $\text{Abs}_{[O]}$ is the oligomycin sample absorbance, Ins is the intercept of the standard curve, and Slp is the slope of the standard curve.

4 Notes

1. Do not worry if the sodium potassium tartrate tetrahydrate does not dissolve totally, for the following addition of sodium hydroxide will guarantee its dissolution.

2. It is vital that from the moment the animal is sacrificed to this point to take as less time as possible (roughly 60–90 s should suffice).

3. Notice that this will release a lot of blood from the liver and, as such, the buffer should be replaced in order to remove it (two to three times for a normal 10 g liver from a 250 g animal should suffice). Take care of not to waste liver material when throwing away the old buffer.

4. The pestle should reach the bottom of the homogenizer, which might not be possible at the first stroke; nevertheless, if the tissue was correctly chopped, at the second stroke it should be possible.

5. This centrifugation allows for the sedimentation of nuclei, red cells, broken and intact cells, and other heavier, unwanted components.

6. Waste a small amount of supernatant at the end of the tube to avoid carryover of pelleted material.

7. To resuspend the mitochondrial pellet use, for example, a watercolor thin paintbrush. The mitochondrial pellet forms a soft brown layer adhered against the tube wall, with a possible dark red central spot, which can be discarded as it consists of pelleted red blood cell contents. It can also have a superficial mobile layer of pelleted mitochondria, which should be discarded (most of it is easily removed when decanting the buffer) as it is formed by damaged mitochondria.

8. 3 or 4 mL of buffer should suffice for a typical 10 g liver.

9. This is important to allow the stabilization of the mitochondrial preparation. Use this time to quantify the mitochondrial preparation. A complete quantification should measure not only protein content but also the activity of citrate synthase, a known mitochondrial marker.

10. Prepare BSA in water, store at –20 °C in aliquots. DOC sodium salt, is also prepared in water, and stored in a light-protected glass bottle. Start by the water, wash buffer and DOC. When all is ready, add the samples to the respective tubes, making at least duplicates. Only when everything is added, add the Biuret reagent.

11. Start by tube 0, and use it to "zero" the spectrophotometer. The r^2 (coefficient of determination) is easily calculated using a standard spreadsheet software and should be between 0.995 and 1 (below 0.995, reject the curve and start again). Using the software again, calculate the intersection of the curve to the y-axis, the curve slope and register the dilution factor (for this example, the dilution factor is 20).

12. The replicates should be roughly similar, i.e., no more than 0.05 absorbance units apart—if they are, reject your tubes and start again.

13. As an example, a slope of 0.1240, and intersection of –0.0113333, a dilution factor of 20 and an average sample absorbance of 0.5110 indicate that the sample has a protein content of 84.2473 mg/mL.

14. This will cause an immediate shift in the absorbance, creating a rate. It is important to record this rate, for it is recommended that only one cuvette be tackled at a time.

15. As before, this will cause an immediate rate shift, for the same precautions are recommended.

16. The tubes with rotenone serve as a negative control and should not demonstrate any increase in absorbance.

17. Reactions should take place at 25 °C, in a light-protected environment, under magnetic stirring, since PMS is a highly light and oxygen-sensitive reagent, for it should be made fresh and used in under a few hours after preparation (or made and kept at –20 °C for a few days, protected from the light). A distinct shift in color from yellow to green from PMS indicates the reagents' decay.

18. A digital oxygen recorder automatically calculates the slope and, as such, indicates the oxygen consumption rate per unit of time. As for analogic measurements, one must know how much oxygen there was inside the chamber, dissolved in the buffer. Since pure water, at 1 atm and 25 °C has roughly 2.06 mM of oxygen atoms dissolved within, it is simple to calculate the slope in terms of time.

19. To the previously described decylubiquinone solution, add a few crystals of lithium borohydrate and mix thoroughly by pipetting, until the solution is clear. If excess crystals are added, a few drops of concentrated HCl should get rid of them. When HCl is added and no bubbling is visible, the solution is ready. This will cause the final pH to be between 2 and 3.

20. Antimycin A is a specific Complex III inhibitor and this will be, as before, the negative control tube.

21. Ascorbate helps maintain TMPD in the reduced state, which supplies electrons to cytochrome c.

22. Reactions should take place at 25 °C, in a light-protected environment, under magnetic stirring. Ascorbate + TMPD can be added simultaneously, but should be kept at –20 °C and light protected, preferably freshly made prior to being used.

23. Use only glass test tubes that have been submerged in an HCl solution (2 M) for at least 24 h. Wash the tubes after the acid bath with ultra-pure water.

References

1. Barrientos A, Fontanesi F, Diaz F (2009) Evaluation of the mitochondrial respiratory chain and oxidative phosphorylation system using polarography and spectrophotometric enzyme assays. Curr Protoc Hum Genet, 19:Unit19–Unit13

2. Gornall AG, Bardawill CJ, David MM (1949) Determination of serum proteins by means of the biuret reaction. J Biol Chem 177(2):751–766

3. Janssen R, Nijtmans LG, van den Heuvel LP, Smeitink JA (2006) Mitochondrial complex I: structure, function and pathology. J Inherit Metab Dis 29(4):499–515

4. Janssen AJM, Trijbels FJM, Sengers RCA, Smeitink JAM, van den Heuvel LP, Wintjes LTM, Stoltenborg-Hogenkamp BJM, Rodenburg RJT (2007) Spectrophotometric assay for complex I of the respiratory chain in tissue samples and cultured fibroblasts. Clin Chem 53(4):729–734

5. Fischer JC, Ruitenbeek W, Trijbels JM, Veerkamp JH, Stadhouders AM, Sengers RC, Janssen AJ (1986) Estimation of NADH oxidation in human skeletal muscle mitochondria. Clin Chim Acta 155(3):263–273

6. Singer TP (1974) Determination of the activity of succinate, NADH, choline, and alpha-glycerophosphate dehydrogenases. Methods Biochem Anal 22:123–175

7. Tisdale HD (1967) Preparation and properties of succinic-cytochrome c reductase (complex II–III). Methods Enzymol 10:213–215

8. Brautigan DL, Ferguson-Miller S, Margoliash E (1978) Mitochondrial cytochrome c: preparation and activity of native and chemically modified cytochromes c. Methods Enzymol 53:128–164

9. Jonckheere AI, Smeitink JAM, Rodenburg RJT (2012) Mitochondrial ATP synthase: architecture, function and pathology. J Inherit Metab Dis 35(2):211–225

Chapter 8

Histoenzymatic Methods for Visualization of the Activity of Individual Mitochondrial Respiratory Chain Complexes in the Muscle Biopsies from Patients with Mitochondrial Defects

Agnieszka Karkucinska-Wieckowska, Maciej Pronicki, and Mariusz R. Wieckowski

Abstract

Investigation of mitochondrial metabolism perturbations and successful diagnosis of patients with mitochondrial abnormalities often requires assessment of human samples like muscle biopsy. Immunohistochemical and histochemical examination of muscle biopsy is an important technique to investigate mitochondrial dysfunction that combined with spectrophotometric and Blue Native electrophoresis techniques can be an important tool to provide diagnosis of mitochondrial disorders. In this chapter we focus on technical description of the methods that are suitable to detect the activity of complex I, II, and IV of mitochondrial respiratory chain in muscle biopsies. The protocols provided can be useful not only for general assessment of mitochondrial activity in studied material, but they are also successfully used in the diagnostic procedures in case of suspicion of mitochondrial disorders.

Key words Mitochondrial respiratory chain complexes, Mitochondrial disorders

1 Introduction

As the center of oxidative metabolism mitochondria are considered as a cellular powerhouse. They produce the majority of cellular ATP but also mitochondria are involved in several other critical metabolic processes [1]. For this reason, and many others, defects in mitochondrial function are often connected with pathological states and can lead to various diseases [2–5]. Special attention should be devoted to the alterations in the mitochondrial respiratory chain. Localization and measurement of changes in the activity of respiratory chain complexes can provide important information about the disease etiology. A wide spectrum of methods enables studying the activity of respiratory chain complexes in isolated mitochondria, intact cells as well as in tissues samples [6–8].

Carlos M. Palmeira and Anabela P. Rolo (eds.), *Mitochondrial Regulation*, Methods in Molecular Biology,
vol. 1241, DOI 10.1007/978-1-4939-1875-1_8, © Springer Science+Business Media New York 2015

Under special experimental conditions the activity of individual mitochondrial respiratory chain complex can be measured separately using spectrophotometric methods based on electron donor/acceptor enzymatic coupled reactions [9]. Blue Native electrophoresis followed by in-gel activity assay can also be successfully applied either for detection of respiratory chain complex deficiency or for visualization deficiencies of the individual respiratory chain complex activity [10, 11]. Histoenzymatic methods presented in our chapter are dedicated for visualization of alterations in the activity of individual mitochondrial respiratory chain complexes in the muscle biopsies from patients with various mitochondrial defects [12–15]. In the case of cytochrome c oxidase (COX) obtained picture is very specific and results in the reduction or absence of COX staining (COX-intermediate and COX-deficient), which indicates abnormalities in complex IV functioning [16]. Here we present examples of normal and pathological pictures of histoenzymatic analysis of complex I, II, and IV in muscle biopsies from healthy donors and patients with mitochondrial defects (*see* **Note 1**).

2 Materials

2.1 Assessment of Cytochrome c Oxidase (COX; Complex IV) Activity in Muscle Biopsies

2.1.1 Solutions and Chemicals

1. Catalase stock: 20 μg/ml (*see* **Note 2**).

2. Formal-calcium (fixer): 15.8 g of calcium chloride dissolved in 60 ml of deionized water and supplemented with 40 ml of 40 % formaldehyde (*see* **Note 3**).

3. 0.1 M phosphate buffer pH 7.4 (*see* **Note 4**): Directly before use 4.05 ml of Na_2HPO_4 solution (1.42 g of Na_2HPO_4 dissolved in 50 ml of deionized water) should be mixed with 0.95 ml of NaH_2PO_4 solution (1.39 g NaH_2PO_4 dissolved in 50 ml of deionized water) and supplemented with 5 ml of deionized water.

4. Incubation solution (to be prepared directly before use): 5 mg of DAB dissolved in 9 ml of 0.1 M phosphate buffer. Next reagents should be added in the following order: 10 mg cytochrome c (*see* **Note 5**), 1 ml of catalase stock, and 750 mg sucrose. Solution should be mixed well without doing foam.

5. Formaldehyde solution for molecular biology 36.5–38 %.

6. Xylene (mixed isomers).

7. Ethanol 96 %.

8. Ethanol anhydrous 99.6 %.

9. DePeX mountant for histology.

2.1.2 Equipment and Accessories

1. Cryostat (Cryotome FSE).

2. Incubator with the ability to achieve below-ambient temperature.

3. Olympus-BX53 with XC50 digital camera and CellSense Dimension image-capture software.

4. Humidity chamber for Immunohistochemical staining.

5. Coplin staining jar.

6. Coated microscope slide Superfrost Plus.

7. Cover glasses.

8. DakoPen.

2.2 Assessment of Succinate Dehydrogenase (SDH; Complex II) Activity in Muscle Biopsies

2.2.1 Solutions and Chemicals

1. NBT stock: 1 mg/ml deionized water (*see* **Note** 7).

2. 1 M Tris–HCl buffer pH 7.4 (*see* **Note 8**): 6.1 g of Tris dissolved in 500 ml of deionized water, supplemented with 37 ml of 1 M HCl solution and adjusted to 1 l with the use of deionized water.

3. Succinate stock (prepared directly before use): 54 mg of sodium succinate dibasic hexahydrate dissolved in 2 ml of deionized water.

4. Incubation solution (prepared directly before use): 2 ml of succinate stock mixed with 1 ml of NBT stock and supplemented with 1 ml of Tris–HCl buffer.

5. Dako faramount aqueous mounting medium.

2.2.2 Equipment and Accessories

1. Cryostat (Cryotome FSE).

2. Incubator with the ability to achieve below-ambient temperature.

3. Olympus-BX53 with XC50 digital camera and CellSense Dimension image-capture software.

4. Humidity chamber for Immunohistochemical staining.

5. Coplin staining jar.

6. Coated microscope slide Superfrost Plus.

7. Cover glasses.

8. DakoPen.

2.3 Assessment of Dehydrogenase NADH (Diaphorase; Complex I) Activity in Muscle Biopsies

2.3.1 Solutions and Chemicals

1. NBT stock: 1 mg/ml deionized water (*see* **Note** 7).

2. 1 M Tris–HCl buffer pH 7.4 (*see* **Note 8**): 6.1 g of Tris dissolved in 500 ml of deionized water, supplemented with 37 ml of 1 M HCl solution and adjusted to 1 l with the use of deionized water.

3. NADH stock (prepared directly before use): 3 mg of β-nicotinamide adenine dinucleotide, reduced, disodium salt (NADH) dissolved in 3 ml of 1 M Tris–HCl buffer.

4. Incubation solution (prepared directly before use): 3 ml of NADH stock mixed with 3 ml of NBT stock solution.

5. Dako faramount aqueous mounting medium.

<table>
<tr><td>

*2.3.2 Equipment
and Accessories*

</td><td>

1. Cryostat (Cryotome FSE).
2. Incubator with the ability to achieve below-ambient temperature.
3. Olympus-BX53 with XC50 digital camera and CellSense Dimension image-capture software.
4. Humidity chamber for Immunohistochemical staining.
5. Coplin staining jar.
6. Coated microscope slide Superfrost Plus.
7. Cover glasses.
8. DakoPen.

</td></tr>
</table>

3   Methods

<table>
<tr><td>

3.1 Muscle Biopsy from Patients Suspected to Carry Mitochondrial Defects

</td><td>

1. Skeletal muscle samples (quadriceps) were obtained by open muscle biopsy from patients suspected to carry mitochondrial defects.
2. Deep muscle biopsies in tissue embedding medium (matrix) and freeze in isopentane cooled by liquid nitrogen.
3. Cut with the use of cryostat from three to five (per coated microscope slide) 8–10 μm sections from transversely orientated muscle blocks and subjected to histoenzymatic staining for COX, SDH and NADH diaphorase activity.

</td></tr>
<tr><td>

3.2 Assessment of COX (Complex IV) Activity in the Muscle Biopsies

</td><td>

In the place where cytochrome c oxidase (complex IV) has an enzymatic activity, DAB molecule is reduced and DAB reduced molecule as a brown precipitate can be observed.

1. Put unfixed cryostat sections on coated microscope slide.
2. With the use of DakoPen outline cryostat sections in order to avoid flow of incubation solution from the microscope slide.
3. Put microscope slides in to the humidity chamber for immunohistochemical staining.
4. Cover outlined cryostat sections with incubation solution and incubate for 1 h at 22 °C.
5. Wash three times with distilled water.
6. Fix in formal-calcium solution in Coplin staining jar for 15 min at room temp.
7. Dehydrate sections in Coplin staining jar containing 96 % ethanol (2×1 min).
8. Next dehydrate sections in Coplin staining jar containing 99.6 % ethanol (2×1 min).
9. Clear in Coplin staining jar containing Xylene (mixed isomers) (3×3 min).

</td></tr>
</table>

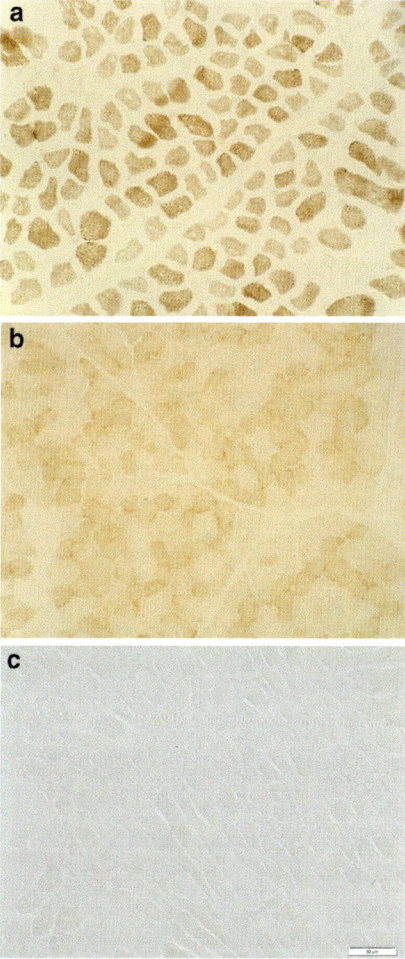

Fig. 1 Histochemical assessment of *cytochrome c oxidase* (*COX; complex IV*) *activity indiagnostic muscle biopsies.* Light microscopy, original magnification 400×. (**a**) Normal COX activity; (**b**) Mosaic COX deficiency suggesting mtDNA heteroplasmy; (**c**) Diffuse total COX deficiency (to be interpreted only together with control positive reaction)

10. Mount in DePeX mountant for histology and cover with cover glass (*see* **Note 6**).

11. Analyze with light microscope and record images with different magnifications—brown reaction product at the sites of active cytochrome oxidase should be observed.

12. For future analysis prepare photographic records from stained sections (Fig. 1).

3.3 Assessment of SDH (Complex II) Activity in the Muscle Biopsies

In the place where SDH (complex II) has an enzymatic activity, NBT molecule is reduced and NBT reduced molecule as a purple formazan deposit with NBT can be observed.

1. Put unfixed cryostat sections on coated microscope slide.

2. With the use of DakoPen outline cryostat sections in order to avoid flow of incubation solution from the microscope slide.

3. Put microscope slides in to the humidity chamber for immunohistochemical staining.

4. Cover outlined cryostat sections with incubation solution and incubate for 90 min at 37 °C.

5. Wash three times with distilled water.

6. Mount in Dako faramount aqueous mounting medium and cover with cover glass (*see* **Note 6**).

7. Analyze with light microscope and record images with different magnifications—bluish-black reaction product at the sites of active succinate dehydrogenase should be observed.

8. For future analysis prepare photographic records from stained sections (Fig. 2).

3.4 Assessment of NADH Diaphorase (Complex I) Activity in the Muscle Biopsies

In the place where NADH diaphorase (complex I) has an enzymatic activity, NBT molecule is reduced and NBT reduced molecule as a purple formazan deposit with NBT can be observed.

1. Put unfixed cryostat sections on coated microscope slide.

2. With the use of DakoPen outline cryostat sections in order to avoid flow of incubation solution from the microscope slide.

3. Put microscope slides in to the humidity chamber for immunohistochemical staining.

4. Cover outlined cryostat sections with incubation solution and incubate for 60 min at 37 °C.

5. Wash three times with distilled water.

6. Mount in Dako faramount aqueous mounting medium and cover with cover glass (*see* **Note 6**).

7. Analyze with light microscope and record images with different magnifications—bluish-black reaction product at the sites of active NADH diaphorase should be observed.

8. For future analysis prepare photographic records from stained sections (Fig. 3).

4   Notes

1. *Ethics*—The studies with the use of skeletal muscle samples (quadriceps) obtained by open muscle biopsy, were carried out in accordance with the Declaration of Helsinki of the World Medical Association and were approved by the Committee of

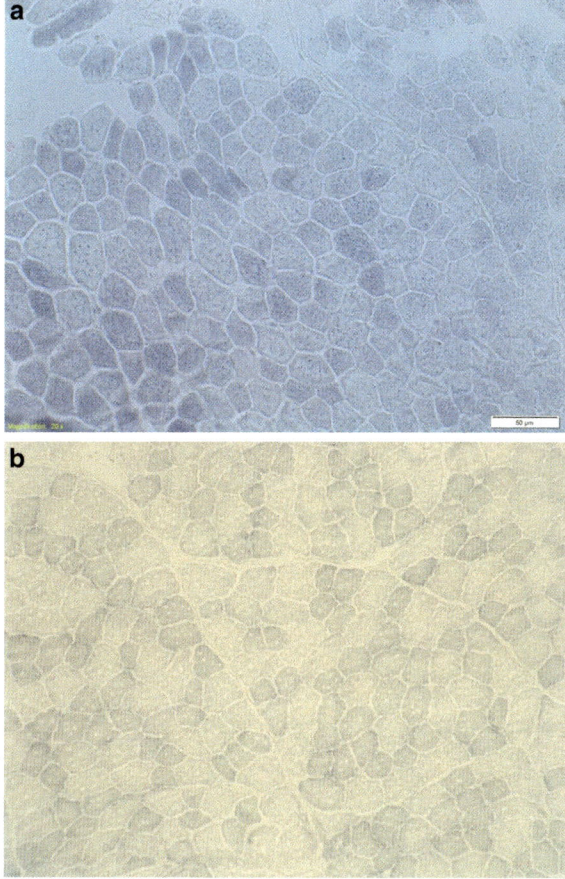

Fig. 2 Histochemical assessment of *SDH (complex II) activity in the diagnostic muscle biopsies*. Light microscopy, original magnification 400×. (**a** and **b**) Different patterns of normal SDH activity

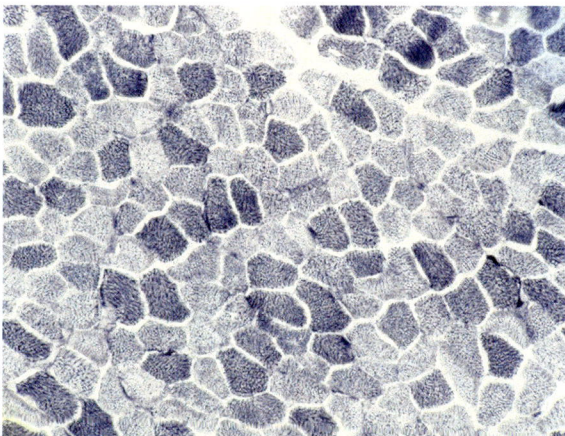

Fig. 3 Histochemical assessment of NADH diaphorase (complex I) activity in a diagnostic human muscle biopsy. Light microscopy, original magnification 400×. Normal pattern of enzymatic activity clearly showing fiber type differentiation. Minimal subsarcolemmal accumulation of reaction product in individual fibers is considered normal in children

Bioethics at The Children's Memorial Health Institute. Informed consent was obtained from the parents before any biopsy or molecular analysis was performed.

2. The stock (approx. 1.2 ml) can be prepared in advance and stored in –20 °C for approx. 1 month.

3. Solution can be prepared in advance and stored in room temperature for approx. 1 month in the dark glass bottle.

4. Na_2HPO_4 solution and NaH_2PO_4 solution can be prepared in advance and stored at 4 °C for approx. 1 month.

5. Use only cytochrome c from horse heart that has been prepared (obtained/purified) without TCA precipitation step. This is very crucial! You can use cyt. c from for example Sigma-Aldrich cat. n. C7752.

6. Mounted sections can be analyzed immediately or up to 1 year.

7. The stock of NBT can be prepared in advance and stored at 4 °C for approx. 1 month.

8. Solution can be prepared in advance and stored at 4 °C for approx. 1 month.

Acknowledgments

This work was supported by the Statutory Funding from Nencki Institute of Experimental Biology, Polish Ministry of Science and Higher Education grant W100/HFSC/2011 and BIO-IMAGing in Research Innovation and Education (FP7-REGPOT-2010-1). Moreover, this work supported by POIG.02.01.00-14-059/09 project titled "Multi-profile modernization of scientific and research infrastructure of The Children's Memorial Health Institute" that was submitted within the frameworks of the competition for Measure 2.1 "Development of high potential research centres", Priority II "R&D Infrastructure" within the frameworks of the Innovative Economy Operational Programme, 2007–2013

References

1. McBride HM, Neuspiel M, Wasiak S (2006) Mitochondria: more than just a powerhouse. Curr Biol 16:R551–R560

2. Lin MT, Beal MF (2006) Mitochondrial dysfunction and oxidative stress in neurodegenerative diseases. Nature 443:787–795

3. Maechler P, Wollheim CB (2001) Mitochondrial function in normal and diabetic beta-cells. Nature 414:807–812

4. McKenzie M, Liolitsa D, Hanna MG (2004) Mitochondrial disease: mutations and mechanisms. Neurochem Res 29:589–600

5. Taylor RW, Turnbull DM (2005) Mitochondrial DNA mutations in human disease. Nat Rev Genet 6:389–402

6. Sciacco M, Bonilla E (1996) Cytochemistry and immunocytochemistry of mitochondria in tissue sections. Methods Enzymol 264:509–521

7. Capaldi RA, Murray J, Byrne L, Janes MS, Marusich MF (2004) Immunological approaches to the characterization and diagnosis of mitochondrial disease. Mitochondrion 4:417–426

8. Pronicki M, Szymańska-Dębińska T, Karkucińska-Więckowska A, Krysiewicz E, Kaczmarewicz E, Bielecka L, Piekutowska-Abramczuk D, Sykut-Cegielska J, Cukrowska B, Więckowski MR (2008) Assessment of respiratory chain function in cultured fibroblasts using cytochemistry, immunocytochemistry and SDS-PAGE. Ann Diagn Pediatr Pathol 12:53–61

9. Kramer KA, Oglesbee D, Hartman SJ, Huey J, Anderson B, Magera MJ, Matern D, Rinaldo P, Robinson BH, Cameron JM, Hahn SH (2005) Automated spectrophotometric analysis of mitochondrial respiratory chain complex enzyme activities in cultured skin fibroblasts. Clin Chem 51(11):2110–2116

10. Karkucińska-Więckowska A, Czajka K, Wasilewski M, Sykut-Cegielska J, Pronicki M, Pronicka E, Zabłocki K, Duszyński J, Więckowski MR (2006) Blue Native Electrophoresis: an additional useful tool to study deficiencies of mitochondrial respiratory chain complexes. Ann Diagn Pediatr Pathol 10(3–4):89–92

11. Lebiedzinska M, Duszynski J, Wieckowski MR (2008) Application of "blue native" electrophoresis in the studies of mitochondrial respiratory chain complexes in physiology and pathology. Postepy Biochem 54:217–223

12. Piekutowska-Abramczuk D, Magner M, Popowska E, Pronicki M, Karczmarewicz E, Sykut-Cegielska J, Kmiec T, Jurkiewicz E, Szymanska-Debinska T, Bielecka L, Krajewska-Walasek M, Vesela K, Zeman J, Pronicka E (2009) SURF1 missense mutations promote a mild Leigh phenotype. Clin Genet 76(2):195–204

13. Pronicki M, Sykut-Cegielska J, Matyja E, Musialowicz J, Karczmarewicz E, Tonska K, Piechota J, Piekutowska-Abramczuk D, Kowalski P, Bartnik E (2007) G8363A mitochondrial DNA mutation is not a rare cause of Leigh syndrome—clinical, biochemical and pathological study of an affected child. Folia Neuropathol 45(4):187–191

14. Pronicki M, Matyja E, Piekutowska-Abramczuk D, Szymanska-Debinska T, Karkucinska-Wieckowska A, Karczmarewicz E, Grajkowska W, Kmiec T, Popowska E, Sykut-Cegielska J (2008) Light and electron microscopy characteristics of the muscle of patients with SURF1 gene mutations associated with Leigh disease. J Clin Pathol 61(4):460–466

15. Seligman AM, Karnovsky MJ, Wasserkrug HJ, Honker JS (1968) Non-droplet ultrastructural demonstration of cytochrome oxidase activity whit polymerizing osmiophilic reagent, DAB. J Cell Biol 38:1

16. Murphy JL, Ratnaike TE, Shang E, Falkous G, Blakely EL, Alston CL, Taivassalo T, Haller RG, Taylor RW, Turnbull DM (2012) Cytochrome c oxidase-intermediate fibres: importance in understanding the pathogenesis and treatment of mitochondrial myopathy. Neuromuscul Disord 22:690–698

Chapter 9

A Method Aimed at Assessing the Functional Consequences of the Supramolecular Organization of the Respiratory Electron Transfer Chain by Time-Resolved Studies

Fabrice Rappaport

Abstract

A steadily increasing number of physiological, biochemical, and structural studies have provided a growing support to the notion that the respiratory electron transfer chain may contain supra-molecular edifices made of the assembly of some, if not all, of its individual links. This structure, usually referred to as the solid state model—in comparison to the liquid state model in which the electron transfer reactions between the membrane bound enzymes are diffusion controlled—is seen as conferring specific kinetic properties to the chain and thus as being highly relevant from a functional point of view. Although the assumption that structural changes are mirrored by functional adjustment is undoubtedly legitimate, experimental evidences supporting it remain scarce. Here we review a recent methodological development aimed at tackling the functional relevance of the supramolecular organization of the respiratory electron transfer chain in intact cells.

Key words Supercomplexes, Time-resolved absorption changes, Kinetic, Electron transfer

1 Introduction

The mitochondria and chloroplast are the power-houses of eukaryotic cells. They are both in charge of the efficient provision of energy by feeding metabolism with its major substrate: ATP. Oxidative-phosphorylation, initially described by Peter Mitchell, relies on the generation, by an electron/proton transfer chain, of an electrochemical transmembrane potential utilized to synthesize ATP; a basic principle that applies to both the respiratory and photosynthetic chain [1]. Beside this functional likeness, these two systems also share ultrastructural characteristics. Indeed, the two electron transfer chains are embedded in a membrane with a strikingly intricate architecture.

Carlos M. Palmeira and Anabela P. Rolo (eds.), *Mitochondrial Regulation*, Methods in Molecular Biology, vol. 1241, DOI 10.1007/978-1-4939-1875-1_9, © Springer Science+Business Media New York 2015

Despite these similarities, the current available knowledge on the function of the photosynthetic chain is more refined than that of its respiratory counterpart and this likely stems from the simple fact that, in the photosynthetic case, the substrate, a photon, can be delivered extremely accurately under the form of light flashes. This constitutes, from an experimental stand-point, a considerable advantage since it allows triggering the whole chain with almost no limitation to the time resolution. Such an approach has proved extremely fruitful since it led to the determination of the kinetic and thermodynamic parameters of most if not all the various electron transfer reactions involved in the photosynthetic process, the time range of which covers ten orders of magnitude. Besides this rare accuracy in the understanding of the function of the various enzymes participating to the photosynthetic process, functional studies have recently pinpointed the importance of the association of these enzymes into "supercomplexes" [2–4] as well as the consequences of the peculiar membrane topology on the function of the chain [5–8]. Although the fundamental principles stated by Mitchell still apply, the emerging picture of the photosynthetic chain clearly diverges from the linear and fluid scheme. As examples, membrane domains strongly restrict the diffusion of electron carriers between membrane bound complexes [9, 7] and supramolecular assembly of various complexes within a single super complex modulates the thermodynamic properties of the components of the chain [2–4]. There is, consequently, a common thinking that the overall properties of the chain cannot be described by the simple combination of the individual properties of its component.

Numerous studies have supported the idea that this also applies to the respiratory chain (*see* Fig. 1, *see* refs. 10, 11 and references therein). Evidences that mitochondrial electron transfer complexes specifically interact to form supramolecular structures come from studies on various types of eukaryotes ranging from unicellular such as yeasts, to plants and mammals (see for example refs. 12–21). Interestingly, these respiratory supercomplexes were shown to sustain higher electron transfer activities in vitro than *in organelle* [12]. Further, mutants in single genes encoding subunits of respiratory chain complexes lead to combined enzyme complex defects [22]. Recently, the functional complementation of two defective respiratory chains by cell fusion have shown that the recovery of the respiratory activity correlates with the abundance of supercomplexes including complexes III and IV, suggesting that supercomplex assembly is a necessary step for respiration [23]. Yet, the understanding of the role of these supramolecular assemblies seems inversely correlated to the number of hypotheses which has been proposed to date: substrate channeling, catalytic enhancement or sequestration of reactive intermediates (*see* ref. 24 for a comprehensive discussion).

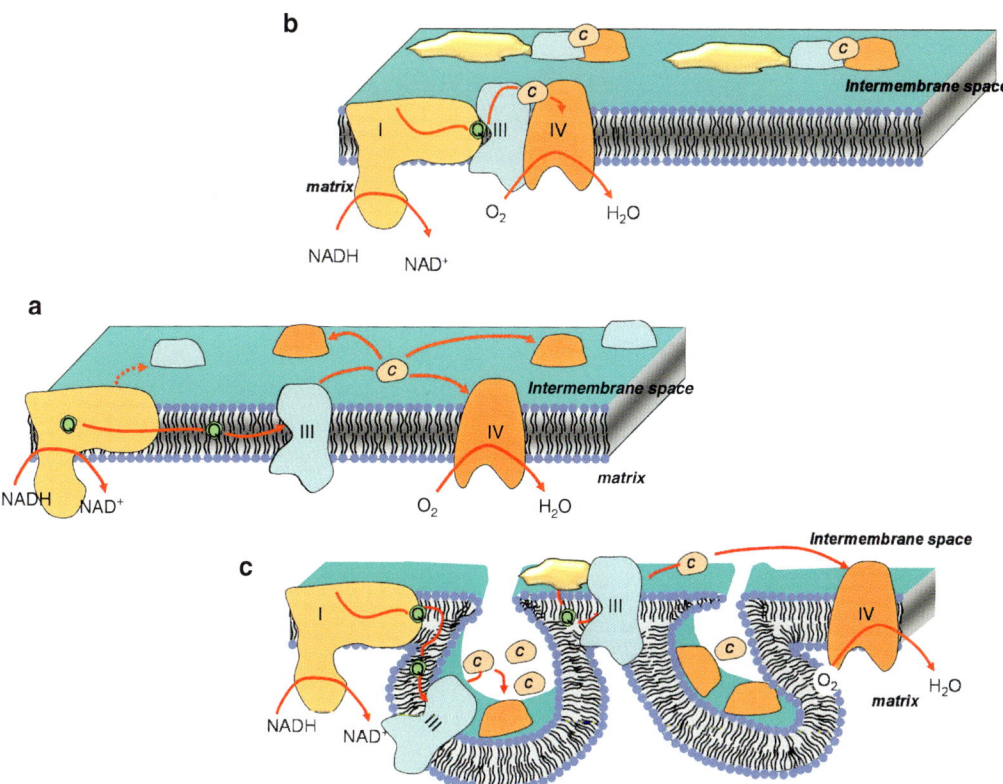

Fig. 1 Mitochondrial respiration involves electron transfer from NADH to molecular oxygen. This overall reaction is mediated by membrane embedded proteins (here depicted as complex I, complex III, and complex IV) and soluble electron carriers, quinones (in *green*) which are membrane soluble and transfer electrons between complexes I and III, and cytochrome *c* which is soluble in the intermembrane space and transfer electrons between complexes III and IV. In panel **a** is shown the "liquid model" in which soluble electron carriers freely diffuse between the membrane embedded complexes. In panel **b**, are shown "respirasomes" or solid state model, in which the membrane complexes are assembled within supramolecular edifices which trap the soluble electron carriers and thereby restrict their diffusion. In panel **c**, is shown the compartmented model, in which the intricate folding of the inner membrane defines local compartments in which cytochrome *c* can freely diffuse but which prevent the equilibration between compartments (taken from the author summary accompanying ref. [35] with permission of the authors)

Supercomplexes are not the only ultra-structural motives susceptible to play an important role in controlling the overall oxidative phosphorylation activity of the respiratory chain in organelles. 3D images provided by EM tomography strongly suggest that, as in the case of the thylakoids in chloroplasts, the internal membrane in mitochondria might be compartmented, and that diffusion between these internal compartments might be restricted. Along these lines, the number and topology of the cristae junction could regulate the rates of ATP production under certain conditions ([25] for a review). Likewise, the shape and volume of the cristae can be expected to affect the diffusion of soluble electron carriers between intra-cristae and intermembrane compartments [26] as in the case of the light-induced swelling of the thylakoid lumen [27]. There is considerable evidence that the mitochondrial inner membrane is a dynamic structure able to change shape rapidly in response to alterations in osmotic or metabolic conditions. If cristae morphology can indeed regulate rates of chemiosmosis, such structural impact may be more than merely passive volume adjustments and could be part of an integral feedback mechanism by which mitochondria respond to environmental perturbations.

The vast majority of experimental data supporting the functional importance of supramolecular structures comes from biochemical and/or structural approaches by which these structures are isolated from their native membranes and then characterized with respect to their biochemical composition and/or structure and to their in vitro activity. The main exceptions to this are studies resorting to measurement of the dependence of the activity of the respiratory chain upon addition of substoichiometric concentration of inhibitors and subsequent flux control analysis [28–33]. Notably, some of these studies contradict earlier conclusions by Hackenbrock et al. that the respiratory electron transfer chain can be described as a fluid system (*see* ref. 34 for a review). To widen the spectrum of the methods available to assess the functional relevance of the structural organization of the mitochondrial membrane and complexes and contribute to this debate by alternative means susceptible to provide a new perspective, we recently developed a different approach aimed at making the respiratory electron transfer chain amenable to time-resolved studies [35]. The rationale for this relies on the fact that, in principle, kinetics studies may allow discriminating readily the transfer of matter through various complexes connected by freely diffusing molecules or sequestered within a single supramolecular edifice. Indeed, whereas the former case involves the diffusion and the encounter of the different actors of the reaction so that their concentrations contribute to determining the rate, in the latter case the reaction takes place within a single molecular entity and is thus described as a (or a sum of) first order process(es) (*see* ref. 24 for a comprehensive discussion of the theoretical aspects).

The main challenge was thus to devise tools to trigger the respiratory process thereby opening the possibility of an accurate analysis of the kinetic and thermodynamic properties of the chain. Such tools do exist that were developed by Britton Chance [36] in the early 1950s but had, since then and to our knowledge, fruitfully but mostly been applied to in vitro studies of complex IV. This approach relies on the respective binding properties of carbon monoxide (CO) and oxygen (O_2) to the catalytic site of complex IV: (1) the rate of release of CO is relatively slow, so that, even in the presence of O_2, the CO-bound state is relatively stable; (2) O_2 binds faster than CO. One can thus take advantage of the possibility to photo-dissociate a bound CO by a laser flash to trigger the release of CO after having mixed the CO-inhibited enzyme with O_2 and, the competition for binding being in favor of O_2, the ensuing turnover of the enzyme [37, 38]. This "flow-flash" technique permitted the accurate kinetic analysis of the various electron and proton transfer reactions associated with the turnover of complex IV (*see* refs. 38–46 for reviews). We have scaled up this approach and applied it to intact eukaryotic cells, following the steps of Britton Chance with an improved time-resolution [35]. Under such conditions, the photo-dissociation of CO should not only trigger the complex IV activity and allow one to follow, in organelles, the cascade of reactions previously characterized in the purified enzyme but also, and in this particular case more importantly, trigger the electron transfer between complex IV and the various complexes located upstream in the chain.

As briefly mentioned above, the methods relies on the light-induced dissociation of CO and the consecutive O_2 binding to Cytochrome oxidase. It thus requires the sample to be conditioned, prior to its light activation, in the presence of a mixture of aqueous CO and O_2. In addition, the aqueous concentration of both gases has to be set with reasonable accuracy, because, the binding of CO or O_2 being second order processes, their respective binding rates depend on their respective concentrations. As regards to the triggering of the respiratory electron transfer chain, these concentrations not only determine the O_2 binding rate but also the respective probability to bind O_2 or CO which are respectively:

$$\rho_{O_2} = \frac{k_{O_2}}{k_{O_2} + k_{CO}} \quad \text{and} \quad \rho_{CO} = \frac{k_{CO}}{k_{O_2} + k_{CO}} \quad \text{where } k_{CO} \text{ and } k_{O2} \text{ are the}$$

pseudo first order rate constant of CO and O_2 binding, respectively.

Because CO is a competitive inhibitor, it does not fully prevent oxygen consumption by the respiratory chain, even in the absence of any light activation. Indeed, the spontaneous dissociation of CO from the reduced heme a_3, even though it is a slow process ($k_{off}^{CO} \sim 0.02\,\text{s}^{-1}$ [47]), allows the binding of oxygen and the subsequent oxidation of a fraction of Cytochrome c Oxidase (CcOx), particularly under relatively high $[O_2]$. This makes it impossible to preconditioned, the sample in the desired state with respect to $[O_2]$ for a long period of time before pulling the light trigger. In principle, the flow-flash approached employed to study in details oxidases (*see* ref. 38 for a review) may solve this issue (see for examples, refs. 36, 37). We resorted to a similar approach yet avoiding the stop-flow steps. As shown in the schematic of the setup (Fig. 2), the sample was preconditioned in a 6–10 mL reservoir connected to a 500 mL bottle, which contains the gas mixture at the desired respective partial pressure, via a peristaltic pump used to flow the gas in and out of the reservoir containing the cell suspension. Strong bubbling insured both the equilibration between the gas

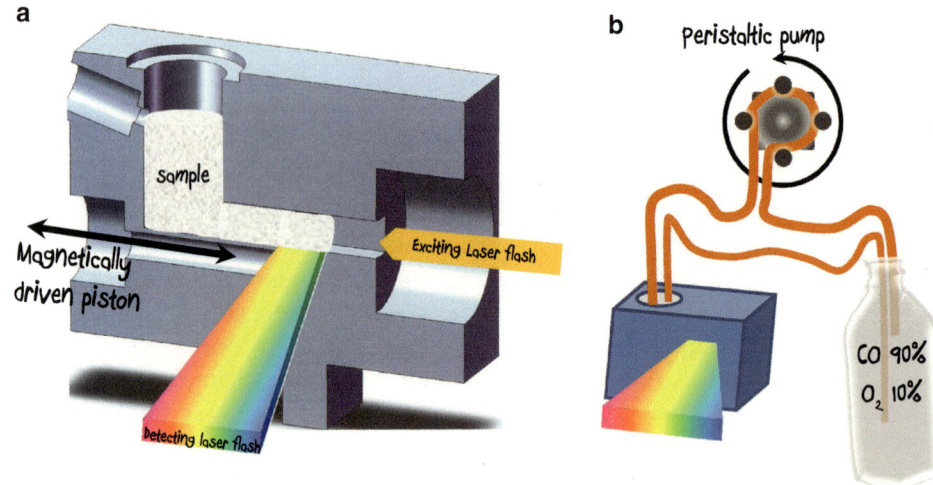

Fig. 2 Schematic illustration of the setup. Panel **a** shows a longitudinal cross-section of the cuvette and sample containing reservoir. The cuvette is an 8 mm side cube, connected to the reservoir by a magnetically driven piston. The reservoir contains up to 10 mL of sample which is preconditioned in terms of gas concentration. To this end, the gas mixture, realized in a 500 mL bottle, is bubble in the reservoir using a peristaltic pump as depicted in Panel **b**

and aqueous phases and stirring of the cell suspension. This reservoir is connected to the cuvette, a cube with 8 mm long sides with three quartz windows, by a piston operated magnetically. The exciting flash is perpendicular to the detecting beam.

The spectrophotometric part of the setup is described in details in [48]. Briefly, detecting and exciting flashes are both provided by OPO resonant cavities (respectively Panther and Slopo, Continuum, Santa Clara, CA, USA) pumped by the third harmonic of pulsed Nd:YAG lasers. The excitation wavelength was 590 or 605 nm. It was filtered out, to avoid saturation of the detecting photodiodes, using cutoff filters (third millennium 570 LP filters, Omega optical, Brattelboro, VT, USA, in combination with 5 mm thick BG39 filters, Schott Optical filters, Mainz, Germany), yet transmitting the detecting probe flashes. When spectral information in the red part of the spectrum was required, excitation was switched to 430 nm, and its intensity was adjusted based on the kinetics and amplitude of the cytochrome c signal at 551 nm.

2   Materials

2.1   Yeasts

The results presented in [35] have been obtained with different strains of *Saccharomyces cerevisiae*, which we elected as our first model system owing to all the obvious advantages this microbe has from an experimental standpoint and to the abundant literature available on the functional and biochemical characteristics of its mitochondrial electron transfer chain. The *Saccharomyces cerevisiae* strains used in this study were derived from BY4742 (from Euroscarf) and from W303-1B [49]. In all strains, the gene *YHB1* encoding a flavohemoprotein was deleted to avoid any possible interference coming from the spectroscopic signal from this enzyme.

However, besides the practical merits of *S. cerevisiae*, there is no strong reason to restrict the application of the approach to this organism. Since our initial work we extended it to other cell systems such as *Yarrowia lipolytica*, another yeast which, at variance with the baker's one, bears an authentic complex I allowing one to assess the existence of functional clusters involving complex I. This is inasmuch important as earlier studies relying on flux control analysis concluded that, the soluble cytochrome c encounters no constraints to its diffusion so that the portion of the chain comprising [Complex III]-[cyt c]-[Complex IV] can be described as a fluid system, quinones, which act as the lipid soluble electron carrier linking Complex I and III, are clustered [10, 30].

2.2 Mammal Cells

To further extend the methods, we also applied it to mammal cells such as RAW 264.7 macrophages. Our preliminary results show that there is no particular limitation to the type of cells which can be used for such studies besides the requirement to be able to suspend the cells in the appropriate liquid medium. Filamentous organisms such as *Podospora anserina* have, until now, proved inappropriate for such studies.

3 Methods

3.1 Culture Medium

Saccharomyces cerevisiae cells were grown on $YPD_{0.5}$ medium, composed of 1 % Yeast Extract, 2 % Peptone, and 0.5 % Glucose. In such a medium, growth displays two distinct phases, the glucose being initially consumed in fermentative pathways, while the ethanol produced during this first phase triggers obligatory respiratory growth once the glucose is depleted [50]. The cells were grown in erlenmeyer flasks (filled at 1/5 with cell culture medium) at 28 °C under vigorous shaking (250 RPM) and harvested in this second exponential growth phase, shortly after the diauxic transition happened, which typically occurred after 10 h. Cells were spun down (1,500×g, 5 min), the pellet was washed in distilled water and, after a second centrifugation, resuspended in MES 40 mM, pH 6.5. Nigericin and valinomycin were added at 10 µM, and the cells were incubated aerobically for at least 10 min.

Yarrowia lipolytica were grown on YPD, 1 % Yeast Extract, 2 % Peptone with no glucose. The cells were grown in erlenmeyer flasks (filled at 1/5 with cell culture medium) at 28 °C under vigorous shaking (250 RPM) and harvested during the logarithmic phase, i.e., between 10 and 15 h of growth. This is important because it has been reported that in the low energy requiring, late stationary-growth phase, complex IV concentration decreases and the cells overexpress NDH2e, an external type II NADH dehydrogenase which, in conjunction with the alternative Oxidase, AOX, bypasses the cytochrome oxidase pathway which is used here to trigger the respiratory chain [51].

3.2 Sample Preparation

One of the critical aspects of the methods is that it relies on the light-induced relief of the inhibition of the respiratory chain which is achieved by the binding of CO to the terminal oxidase. To be effective, this binding must indeed occur and this implies keeping the terminal oxidase in its reduced state. Under the steady state conditions of the preconditioned sample, a reasonable approximation of the fraction of oxidase in the desired "reduced and CO bound" state is $\dfrac{1}{1+\dfrac{k_{\mathrm{off}}^{\mathrm{CO}}.k_{\mathrm{on}}^{\mathrm{O_2}}.[O_2]}{k^{\mathrm{red}}.k_{\mathrm{on}}^{\mathrm{CO}}.[CO]}}$ where $k_{\mathrm{off}}^{\mathrm{CO}}$ is the rate of spon-

taneous dissociation of CO from its binding site ($k_{\mathrm{off}}^{\mathrm{CO}} \sim 0.02\,\mathrm{s^{-1}}$ [47]), $k_{\mathrm{on}}^{\mathrm{O_2}}$ ($\sim 9 \times 10^7\,\mathrm{M^{-1}\,s^{-1}}$, [37, 52–54]) and $k_{\mathrm{on}}^{\mathrm{CO}}$ ($6 \times 10^4\,\mathrm{M^{-1}\,s^{-1}}$, [37, 54, 47]) are the bimolecular rate constants of oxygen and CO binding, respectively, and k^{red} is the overall reduction rate of the enzyme.

In a typical experiment, the partial pressure for CO is ~1 atm which translates into an aqueous concentration of ~900 µM and the concentration of aqueous oxygen is in the 20–150 µM range. Under these conditions $\dfrac{k_{\mathrm{off}}^{\mathrm{CO}}.k_{\mathrm{on}}^{\mathrm{O_2}}.[O_2]}{k_{\mathrm{on}}^{\mathrm{CO}}.[CO]}$ is in the 0.6–5 s^{-1} range so that k^{red} must lie in the 6–50 s^{-1} per oxidase range to insure at least 91 % of the oxidase in the photo-activatable state. This, rough, estimate is commensurable to the overall flux of a reasonably active electron transfer respiratory chain thus showing that the appropriate experimental conditions can be achieved. Notably, it can prove worthwhile, from an experimental standpoint, to add uncouplers such as valinomycin (1 µM), gramicidin (1 µM), or nigericin (which is not strictly speaking an uncoupler but a H$^+$/K$^+$ exchanger) in order to relieve the thermodynamic pressure exerted by the proton motive force and set k^{red} to its maximal value. In addition the above estimate shows that any impairment or inhibition of the respiratory chain upstream of the oxidase will inevitably result in the pre-oxidation of the chain unless [O$_2$] is maintained low enough to keep the $\dfrac{k_{\mathrm{off}}^{\mathrm{CO}}.k_{\mathrm{on}}^{\mathrm{O_2}}.[O_2]}{k^{\mathrm{red}}.k_{\mathrm{on}}^{\mathrm{CO}}.[CO]}$ much lower than 1.

4   Expected Insights Specific to the Approach

At this stage one may still wonder whether the insights that can be expected from the present approach are worth the methodological investment. At first sight, it barely improves, by several of order of magnitude, the time resolution of the kinetic analysis of the respiratory chain. Even though it sounds as a methodological feat, it would remain of little added value if it was nothing more than that. But is it?

As we will discuss below an accurate kinetic analysis is a powerful tool to characterize the supramolecular organization of bioenergetic membranes. This simply stems from the fact that this organization, i.e., clustering of soluble electron carrier or molecular packing owing to high protein density in membranes or membrane compartmentalization etc., will have kinetic consequences. The most straightforward illustration of this statement is that intra-complex electron transfer reactions are first order reactions, whereas inter-complexes reactions are second order processes. In the following we briefly review the various kinetic characteristics that can be exploited to gain insights into these issues and how they can be addressed from an experimental standpoint.

4.1   The Kinetic Signature of Prebound Soluble Carriers

Electron transfer rates steeply decrease as the distance between the electron donor and acceptor increases. As a consequence, electron transfer reaction between two distinct proteins require they encounter and the subsequent formation of a complex which is long-lived enough to allow the redox reaction to occur. Thus electron transfer reactions within a preformed complex are significantly faster than those involving this diffusion limiting step as documented in details in the photosynthetic case of electron transfer between Photosystem I and plastocyanin [55–58]. In the case of Cytochrome oxidase and cytochrome c case electron transfer from reduced, bound cytochrome c to the oxidized Cu_A has been shown to occur in vitro with rates ranging from 7×10^3 to 6×10^4 s^{-1} [59, 60]. As another example, in bacterial caa_3-type oxidases which contain a c-type cytochrome subunit, the oxidation of heme c following photoactivation develops in the 30–240 μs range [61–63]. Thus, a stable complex with a prebound cytochrome c should be evident as a fast kinetic component, as illustrated in vivo by Trouillard et al. [35].

4.2   Distinguishing First and Second Order Processes

By definition, second order processes differ from first order ones by the dependence of their kinetics on the concentration of the substrate. This provides straightforward experimental grounds to discriminate mechanism involving diffusion and encounter of the reactants (second order) from intra-complex reactions (first order), by tuning the relative stoichiometry of the interactants.

4.2.1 Genetic Control of the Relative Stoichiometry

Most of the cellular models mentioned above are amenable to genetic approaches that potentially allow tuning the relative abundance of the interacting partners (*see* ref. 64 for a "photosynthetic example"). As an example, we compared two different *S. cerevisiae* strains expressing two different cytochrome c isoforms having different affinity for cytochrome oxidase. The comparison of the kinetics of oxidation of cytochrome *c* in the two strains allowed us to conclude that the formation of [cytochrome *c*]-[Cytochrome oxidase] complexes was essentially controlled by the affinity of one for the other and not but a particular clustering that would sequester cytochrome c and prevent it from freely diffusing [35]. Alternatively, RNA$_i$ knocked-down mutants may prove a powerful tool to manipulate the relative stoichiometric ratio between different complexes in the chain.

4.2.2 Controlling the Fraction of Photo-Activated Cytochrome Oxidase

The substrate of the CO photo-dissociation event, which triggers the sequence of electron transfer, being photon, its delivery can be tuned accurately by simply changing the intensity of the triggering laser flash (see supplementary S1 in [35]) and the fraction of cytochrome oxidase effectively undergoing photo-dissociation can be directly measured using the specific absorbance changes at 445 nm. The fraction of photo-dissociated, and thus potentially oxidized (see below), Cytochrome oxidase can thus be readily controlled while keeping the initial amount of reduced cytochrome *c* constant, thereby opening the possibility to assess the issue as to whether the oxidase has access to a pool of soluble and freely diffusing cytochrome *c*, or, on the contrary, to a limited and fixed number of bound and sequestered cytochrome c.

4.2.3 Controlling the Number of Light-Induced Turnovers

CO is a competitive inhibitor that binds to the reduced state of the a_3 heme of cytochrome c oxidase. When it competes for binding with another gas, such as oxygen, the two different events can thus considered as independent events so that their respective binding probabilities are: $\rho_{CO} = \dfrac{k_{CO}}{k_{CO} + k_{O_2}}, \rho_{O_2} = \dfrac{k_{O_2}}{k_{CO} + k_{O_2}}$, where k_{O2} and k_{CO} are the binding rates of CO and O_2. These two rates can be experimentally controlled by changing the aqueous concentration of the gases or, equivalently, their respective partial pressure in the gas mixture. Two extreme cases can be considered to illustrate the practical interest of this possibility to tune the competition for binding. The bimolecular rate constant of oxygen binding is about three orders of magnitude larger than that of CO binding. Since the respective solubility of the two gases are similar, a tenfold excess of CO with respect to oxygen warrants that the binding of the substrate outcompete that of CO with a probability of ~0.99 and occur in the tens of microseconds time range. Under such conditions, the rate of oxygen binding is commensurate to cytochrome c oxidation when it is prebound to the oxidase and faster when the electron transfer involves the diffusion limited encounter of the two partners [35], thereby enabling the characterization of the interaction between the membrane bound complex with its soluble electron donor. The other extreme case is that of a low partial pressure in oxygen. At first sight such a case may seem of little practical merit since the oxygen binding rate (and thus the time resolution of the methods) decreases linearly with oxygen concentration. Thus, at low oxygen concentration the binding is so slow that it becomes rate limiting and these precludes any kinetic insights into the downstream reactions. However, decreasing the oxygen binding rate also diminishes its binding probability and this opens the possibility of single turnover experiments.

Once photo-activated, the fraction of CcOx having bound O_2, ρ_{O_2}, may undergo multiple turnovers as the competition between CO and O_2 continues until CO binds to the reduced heme a_3. These successive turnovers of one photo-oxidized CcOx thus constitute a series of independent events where O_2 competes with CO for binding, with the probability ρ_{O_2}. This corresponds to a sequential repeat of independent Bernoulli trials, i.e., an experiment with only two possible random outcomes, here O_2 or CO binding. Quantitatively, the sequence is as follows: ρ_{O_2} of the photoactivated CcOx bind O_2 rather than CO, $\rho_{O_2}{}^2$ undergo a second turnover, $\rho_{O_2}{}^3$ a third one, etc. In this framework, which only assumes that each binding event, either O_2 or Co, is independent from the preceding ones, the number of turnovers occurring before CO binds is $\gamma = \dfrac{1}{\rho_{CO}} - 1 = \dfrac{k_{O_2}.[O_2]}{k_{CO}.[CO]}$. This equation shows that the model predicts a linear relationship between the number of turnovers occurring before CO rebinds and the concentration of O_2. As shown in Fig. 3, this relationship was verified in the case of *S. cerevisiae* when using the lifetime of the oxidized a-a_3 hemes as a measure of the number of turnovers undergone by the cytochrome oxidases.

On average, when $\rho_{O_2} \ll 1$ each photo-oxidized CcOx thus undergo slightly more than a single-turnover, and this provides a simple mean to tune the number of oxidizing equivalents injected in the respiratory chain. In such conditions, where a limited fraction of cytochrome *c* is oxidized by the light-induced turnovers of CcOx, one can address the issue as to whether the probability to oxidize a given cytochrome *c* is independent of the redox state of the other cytochromes *c* in the pool, as expected in a fully stochastic system with no diffusion constraints; or, alternatively, if this probability evolves as the cytochrome *c* pool gets progressively oxidized as a result of the existence of a wide distribution in the size of the cytochrome *c* pool accessible to the light-induced oxidation burst [35].

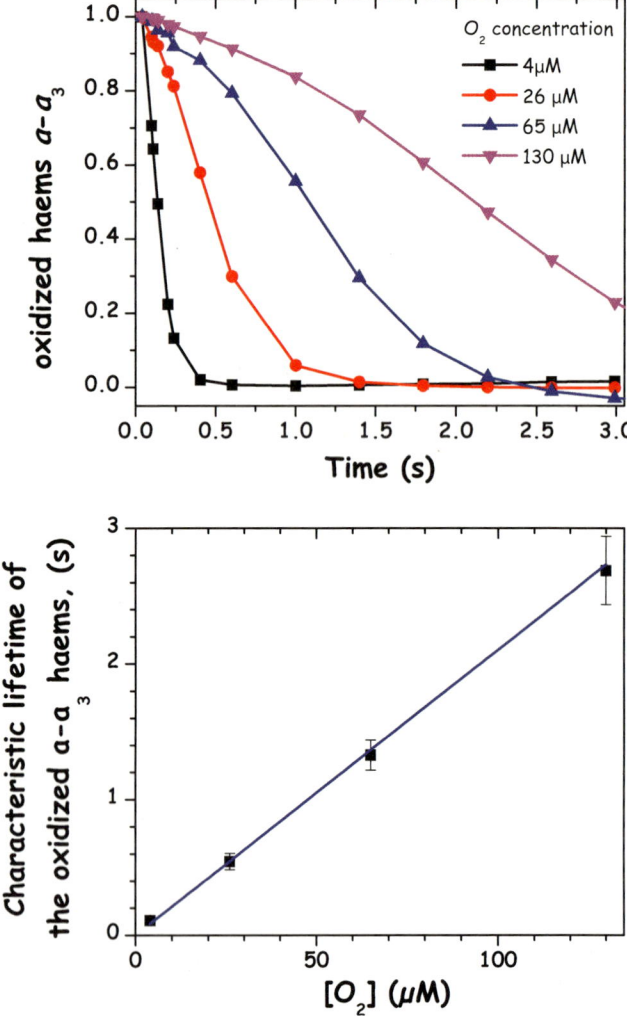

Fig. 3 Dependence of the number of turnovers undergone by the cytochrome oxidase upon oxygen concentration in *S. cerevisiae*. The *top panel* shows the transient oxidation and re-reduction of the *a-a*3 hemes with various oxygen concentrations. The lifetime increases with oxygen concentration and this reflects the competition for binding to the reduced binuclear center between molecular oxygen and CO. As developed in the text, assuming the binding events to each single oxidase are independent from one another the probability law is a simple geometric law which predicts a linear dependence of the number of turnovers upon oxygen concentration. As shown in the *bottom panel*, this expectation is satisfyingly met in the case of *S. cerevisiae*

Acknowledgements

Warm acknowledgements are due to Daniel Béal and Martin Trouillard without whom the photo-activated respiratory electron transfer chain would have remained a project. This work was supported by CNRS, by the ANR (ANR-07-BLAN-0360-01), and by the "Initiative d'Excellence" program from the French State (Grant "DYNAMO", ANR-11-LABX-0011-01).

References

1. Mitchell P (1966) Chemiosmotic coupling in oxidative and photosynthetic phosphorylation. Biol Rev Camb Philos Soc 41:445–502

2. Iwai M, Takizawa K, Tokutsu R, Okamuro A, Takahashi Y, Minagawa J (2010) Isolation of the elusive supercomplex that drives cyclic electron flow in photosynthesis. Nature 464:1210–1213

3. Terashima M, Petroutsos D, Hudig M, Tolstygina I, Trompelt K, Gabelein P, Fufezan C, Kudla J, Weinl S, Finazzi G, Hippler M (2012) Calcium-dependent regulation of cyclic photosynthetic electron transfer by a CAS, ANR1, and PGRL1 complex. Proc Natl Acad Sci U S A 109:17717–17722

4. Takahashi H, Clowez S, Wollman FA, Vallon O, Rappaport F (2013) Cyclic electron flow is redox-controlled but independent of state transition. Nat Commun 4:1954

5. Vermeglio A, Joliot P (1999) The photosynthetic apparatus of Rhodobacter sphaeroides. Trends Microbiol 7:435–440

6. Comayras F, Jungas C, Lavergne J (2005) Functional consequences of the organization of the photosynthetic apparatus in Rhodobacter sphaeroides. I. Quinone domains and excitation transfer in chromatophores and reaction center antenna complexes. J Biol Chem 280:11203–11213

7. Kirchhoff H (2008) Molecular crowding and order in photosynthetic membranes. Trends Plant Sci 13:201–207

8. Joliot P, Johnson GN (2011) Regulation of cyclic and linear electron flow in higher plants. Proc Natl Acad Sci U S A 108:13317–13322

9. Lavergne J, Joliot P (1991) Restricted diffusion in photosynthetic membranes. Trends Biochem Sci 16:129–134

10. Lenaz G, Genova ML (2009) Mobility and function of coenzyme Q (ubiquinone) in the mitochondrial respiratory chain. Biochim Biophys Acta 1787:563–573

11. Genova ML, Lenaz G (2013) A critical appraisal of the role of respiratory supercomplexes in mitochondria. Biol Chem 394:631–639

12. Schagger H, Pfeiffer K (2000) Supercomplexes in the respiratory chains of yeast and mammalian mitochondria. EMBO J 19:1777–1783

13. Schagger H (2002) Respiratory chain supercomplexes of mitochondria and bacteria. Biochim Biophys Acta 1555:154–159

14. Krause F, Reifschneider NH, Vocke D, Seelert H, Rexroth S, Dencher NA (2004) "Respirasome"-like supercomplexes in green leaf mitochondria of spinach. J Biol Chem 279:48369–48375

15. Dudkina NV, Heinemeyer J, Sunderhaus S, Boekema EJ, Braun HP (2006) Respiratory chain supercomplexes in the plant mitochondrial membrane. Trends Plant Sci 11:232–240

16. Wittig I, Carrozzo R, Santorelli FM, Schagger H (2006) Supercomplexes and subcomplexes of mitochondrial oxidative phosphorylation. Biochim Biophys Acta 1757:1066–1072

17. Acin-Perez R, Fernandez-Silva P, Peleato ML, Perez-Martos A, Enriquez JA (2008) Respiratory active mitochondrial supercomplexes. Mol Cell 32:529–539

18. Bultema JB, Braun HP, Boekema EJ, Kouril R (2009) Megacomplex organization of the oxidative phosphorylation system by structural analysis of respiratory supercomplexes from potato. Biochim Biophys Acta 1787:60–67

19. Dudkina NV, Kouril R, Peters K, Braun HP, Boekema EJ (2010) Structure and function of mitochondrial supercomplexes. Biochim Biophys Acta 1797:664–670

20. Althoff T, Mills DJ, Popot JL, Kuhlbrandt W (2011) Arrangement of electron transport chain components in bovine mitochondrial supercomplex I(1)III(2)IV(1). EMBO J 30(22):4652–4664

21. Lapuente-Brun E, Moreno-Loshuertos R, Acin-Perez R, Latorre-Pellicer A, Colas C, Balsa E, Perales-Clemente E, Quiros PM, Calvo E, Rodriguez-Hernandez MA, Navas P, Cruz R, Carracedo A, Lopez-Otin C, Perez-Martos A, Fernandez-Silva P, Fernandez-Vizarra E, Enriquez JA (2013) Supercomplex assembly determines electron flux in the mitochondrial electron transport chain. Science 340:1567–1570

22. Grad LI, Lemire BD (2004) Mitochondrial complex I mutations in Caenorhabditis elegans produce cytochrome c oxidase deficiency, oxidative stress and vitamin-responsive lactic acidosis. Hum Mol Genet 13:303–314

23. D'Aurelio M, Gajewski CD, Lenaz G, Manfredi G (2006) Respiratory chain supercomplexes set the threshold for respiration defects in human mtDNA mutant cybrids. Hum Mol Genet 15:2157–2169

24. Lavergne J (2009) Clustering of electron transfer components: kinetic and thermodynamic consequences. In: Laisk A, Nedbal L, Govindjee (eds) Understanding complexity from molecules to ecosystems, vol 29. Springer, Dordrecht, pp 177–205

25. Frey TG, Mannella CA (2000) The internal structure of mitochondria. Trends Biochem Sci 25:319–324

26. Davies KM, Anselmi C, Wittig I, Faraldo-Gomez JD, Kuhlbrandt W (2012) Structure of the yeast F1Fo-ATP synthase dimer and its role in shaping the mitochondrial cristae. Proc Natl Acad Sci U S A 109:13602–13607

27. Kirchhoff H, Hall C, Wood M, Herbstova M, Tsabari O, Nevo R, Charuvi D, Shimoni E, Reich Z (2011) Dynamic control of protein diffusion within the granal thylakoid lumen. Proc Natl Acad Sci U S A 108:20248–20253

28. Boumans H, vanGaalen MCM, Grivell LA, Berden JA (1997) Differential inhibition of the yeast bc(1) complex by phenanthrolines and ferroin—implications for structure and catalytic mechanism. J Biol Chem 272:16753–16760

29. Boumans H, Grivell LA, Berden JA (1998) The respiratory chain in yeast behaves as a single functional unit. J Biol Chem 273:4872–4877

30. Bianchi C, Genova ML, Parenti Castelli G, Lenaz G (2004) The mitochondrial respiratory chain is partially organized in a supercomplex assembly: kinetic evidence using flux control analysis. J Biol Chem 279:36562–36569

31. Lenaz G, Genova ML (2007) Kinetics of integrated electron transfer in the mitochondrial respiratory chain: random collisions vs. solid state electron channeling. Am J Physiol Cell Physiol 292:C1221–C1239

32. Kaambre T, Chekulayev V, Shevchuk I, Karu-Varikmaa M, Timohhina N, Tepp K, Bogovskaja J, Kutner R, Valvere V, Saks V (2012) Metabolic control analysis of cellular respiration in situ in intraoperational samples of human breast cancer. J Bioenerg Biomembr 44:539–558

33. Kaambre T, Chekulayev V, Shevchuk I, Tepp K, Timohhina N, Varikmaa M, Bagur R, Klepinin A, Anmann T, Koit A, Kaldma A, Guzun R, Valvere V, Saks V (2013) Metabolic control analysis of respiration in human cancer tissue. Front Physiol 4:151

34. Hackenbrock CR, Chazotte B, Gupte SS (1986) The random collision model and a critical assessment of diffusion and collision in mitochondrial electron transport. J Bioenerg Biomembr 18:331–368

35. Trouillard M, Meunier B, Rappaport F (2011) Questioning the functional relevance of mitochondrial supercomplexes by time-resolved analysis of the respiratory chain. Proc Natl Acad Sci U S A 108:E1027–E1034

36. Chance B (1954) Spectrophotometry of intracellular respiratory pigments. Science 120:767–775

37. Gibson QH, Greenwood C (1963) Reactions of cytochrome oxidase with oxygen and carbon monoxide. Biochem J 86:541–554

38. Babcock GT, Wikstrom M (1992) Oxygen activation and the conservation of energy in cell respiration. Nature 356:301–309

39. Einarsdottir O, Szundi I (2004) Time-resolved optical absorption studies of cytochrome oxidase dynamics. Biochim Biophys Acta 1655:263–273

40. Brzezinski P, Adelroth P (2006) Design principles of proton-pumping haem-copper oxidases. Curr Opin Struct Biol 16:465–472

41. Wikstrom M, Verkhovsky MI (2006) Towards the mechanism of proton pumping by the haem-copper oxidases. Biochim Biophys Acta 1757:1047–1051

42. Belevich I, Verkhovsky MI (2008) Molecular mechanism of proton translocation by cytochrome c oxidase. Antioxid Redox Signal 10:1–29

43. Brzezinski P, Gennis RB (2008) Cytochrome c oxidase: exciting progress and remaining mysteries. J Bioenerg Biomembr 40:521–531

44. Kaila VR, Verkhovsky MI, Wikstrom M (2010) Proton-coupled electron transfer in cytochrome oxidase. Chem Rev 110:7062–7081

45. von Ballmoos C, Gennis RB, Adelroth P, Brzezinski P (2011) Kinetic design of the respiratory oxidases. Proc Natl Acad Sci U S A 108:11057–11062

46. von Ballmoos C, Adelroth P, Gennis RB, Brzezinski P (2012) Proton transfer in ba(3) cytochrome c oxidase from Thermus thermophilus. Biochim Biophys Acta 1817:650–657

47. Cooper CE, Brown GC (2008) The inhibition of mitochondrial cytochrome oxidase by the gases carbon monoxide, nitric oxide, hydrogen cyanide and hydrogen sulfide: chemical mechanism and physiological significance. J Bioenerg Biomembr 40:533–539

48. Beal D, Rappaport F, Joliot P (1999) A new high-sensitivity 10-ns time-resolution spectrophotometric technique adapted to in vivo analysis of the photosynthetic apparatus. Rev Sci Instrum 70:202–207

49. Thomas BJ, Rothstein R (1989) Elevated recombination rates in transcriptionally active DNA. Cell 56:619–630

50. Rigoulet M, Mourier A, Galinier A, Casteilla L, Devin A (2010) Electron competition process

in respiratory chain: regulatory mechanisms and physiological functions. Biochim Biophys Acta 1797:671–677

51. Guerrero-Castillo S, Cabrera-Orefice A, Vazquez-Acevedo M, Gonzalez-Halphen D, Uribe-Carvajal S (2012) During the stationary growth phase, Yarrowia lipolytica prevents the overproduction of reactive oxygen species by activating an uncoupled mitochondrial respiratory pathway. Biochim Biophys Acta 1817:353–362

52. Oliveberg M, Brzezinski P, Malmstrom BG (1989) The effect of pH and temperature on the reaction of fully reduced and mixed-valence cytochrome c oxidase with dioxygen. Biochim Biophys Acta 977:322–328

53. Verkhovsky MI, Morgan JE, Wikstrom M (1994) Oxygen binding and activation: early steps in the reaction of oxygen with cytochrome c oxidase. Biochemistry 33:3079–3086

54. Brunori M, Giuffre A, Sarti P (2005) Cytochrome c oxidase, ligands and electrons. J Inorg Biochem 99:324–336

55. Farah J, Rappaport F, Choquet Y, Joliot P, Rochaix JD (1995) Isolation of a psaf-deficient mutant of Chlamydomonas reinhardtii: efficient interaction of plastocyanin with the photosystem I reaction center is mediated by the psaf subunit. EMBO J 14:4976–4984, 4971

56. Hippler M, Drepper F, Farah J, Rochaix JD (1997) Fast electron transfer from cytochrome c6 and plastocyanin to photosystem I of Chlamydomonas reinhardtii requires PsaF. Biochemistry 36:6343–6349

57. Hippler M, Drepper F (2006) Electron transfer between Photosystem I and plastocyanin or cytochrome c_6. In: Golbeck J (ed) Photosystem

I: the light-driven plastocyanin ferredoxin oxidoreductase. Kluwer, Dordrecht, pp 499–513

58. Santabarbara S, Redding KE, Rappaport F (2009) Temperature dependence of the reduction of P(700)(+) by tightly bound plastocyanin in vivo. Biochemistry 48:10457–10466

59. Hill BC (1991) The reaction of the electrostatic cytochrome c-cytochrome oxidase complex with oxygen. J Biol Chem 266:2219–2226

60. Geren LM, Beasley JR, Fine BR, Saunders AJ, Hibdon S, Pielak GJ, Durham B, Millett F (1995) Design of a ruthenium-cytochrome c derivative to measure electron transfer to the initial acceptor in cytochrome c oxidase. J Biol Chem 270:2466–2472

61. Hirota S, Svensson-Ek M, Adelroth P, Sone N, Nilsson T, Malmstrom BG, Brzezinski P (1996) A flash-photolysis study of the reactions of a caa3-type cytochrome oxidase with dioxygen and carbon monoxide. J Bioenerg Biomembr 28:495–501

62. Brzezinski P, Wilson MT (1997) Photochemical electron injection into redox-active proteins. Proc Natl Acad Sci U S A 94:6176–6179

63. Sigurdson H, Namslauer A, Pereira MM, Teixeira M, Brzezinski P (2001) Ligand binding and the catalytic reaction of cytochrome caa(3) from the thermophilic bacterium Rhodothermus marinus. Biochemistry 40:10578–10585

64. Pesaresi P, Scharfenberg M, Weigel M, Granlund I, Schroder WP, Finazzi G, Rappaport F, Masiero S, Furini A, Jahns P, Leister D (2009) Mutants, overexpressors, and interactors of Arabidopsis plastocyanin isoforms: revised roles of plastocyanin in photosynthetic electron flow and thylakoid redox state. Mol Plant 2:236–248

Chapter 10

BN-PAGE-Based Approach to Study Thyroid Hormones and Mitochondrial Function

Elena Silvestri, Assunta Lombardi, Federica Cioffi, and Fernando Goglia

Abstract

In recent years, a number of advancements have been made in the study of entire mitochondrial proteomes in both physiological and pathological conditions. Naturally occurring iodothyronines (i.e., T3 and T2) greatly influence mitochondrial oxidative capacity, directly or indirectly affecting the structure and function of the respiratory chain components. Blue native PAGE (BN-PAGE) can be used to isolate enzymatically active oxidative phosphorylation (OXPHOS) complexes in one step, allowing the clinical diagnosis of mitochondrial metabolism by monitoring OXPHOS catalytic and/or structural features. Protocols for isolating mammalian liver mitochondria and subsequent one-dimensional (1D) BN-PAGE will be described in relation to the impact of thyroid hormones on mitochondrial bioenergetics.

Key words Thyroid hormone, Iodothyronine, Mitochondrion, Respiratory chain, BN-PAGE

1 Introduction

Mitochondria are "hybrid" and dynamic organelles resulting from the coordinated expression of both the nuclear and their own genome [1, 2], critically relevant for energy homeostasis, metabolism, regulation of apoptosis, and proper cell viability. Physiologically, mitochondrial functions are ensured and orchestrated by metabolic (i.e., hormones), environmental, and developmental signals, allowing tissues to adjust their energy production according to changing demands. Understanding how mitochondria follow these adjustments would provide invaluable information and give insight into both mitochondrial functions and mitochondria-associated diseases. Importantly, mitochondria are key subcellular targets for thyroid hormones [THs; thyroxine (T4) and $3,3',5$-triiodo-L-thyronine (T3)] and iodothyronines [i.e., $3,5$-diiodo-L-thyronine (T2)] that regulate energy and substrate utilization, which are closely dependent on their effects on mitochondrial functions [3, 4]. Indeed, extensive changes occur in the mitochondrial compartment in response either to THs or

Carlos M. Palmeira and Anabela P. Rolo (eds.), *Mitochondrial Regulation*, Methods in Molecular Biology, vol. 1241, DOI 10.1007/978-1-4939-1875-1_10, © Springer Science+Business Media New York 2015

to physiological/pathological states involving changes in the activity of the thyroid gland [4, 5]. The impact of THs on mitochondrial function is particularly evident in metabolically active tissues, including skeletal muscle, heart, kidney, and liver. The liver, in particular, has a central role in the potent hypolipidemic effect of both T3 and T2 [6–8], which exert a strong inhibitory effect on the development of steatosis [6, 9, 10]. This antisteatotic effect implies a facilitation of fatty acid transport, an induction of fatty acid oxidation, and a reduction of the severity of liver injury determined by the serum levels of transaminases, suggesting a common mechanism of mitochondrial metabolism modulation [6, 9, 10].

To study the response of mitochondria to THs under physiological, pathological, and pharmacological conditions, an integrated multidisciplinary approach is required [11]. Indeed, the high level of compartmentalization in mitochondria and the existence of multipolypeptide complexes that contain hydrophobic proteins in close contact with the membrane lipids, peripheral proteins, and nonprotein cofactors imply that a deep structural/functional study of the mitoproteome requires an appropriate combination of different tools to compensate for the limits imposed by each individual technique [11]. During the last decade, proteomics has been increasingly used to identify and quantify mitochondrial proteins related to cellular perturbations, enforcing data from metabolites and gene sequences both in physiological and pathological situations [12]. To study the effects of iodothyronine administration on total tissue and subcellular compartments in metabolically active tissues, we performed high-resolution differential proteomic analyzes combining two-dimensional gel electrophoresis (2D-E) and subsequent matrix-assisted laser desorption/ionization time-of-flight mass spectrometry (MALDI-TOF MS) and nano liquid chromatography-electrospray ionization-linear ion trap (LC-ESI-LIT)-MS/MS techniques [7, 13–15]. To gain deeper insights into both the mitochondrial response mechanisms to iodothyronines and the resulting proteome alterations, we employed one-dimensional (1D) blue native PAGE [1D BN-PAGE] to examine the abundance of mitochondrial OXPHOS complexes as well as their supramolecular organization and activity [7, 16, 17]. In particular, BN-PAGE was used to separate mitochondrial proteins and complexes in the mass range of 10 kDa to 10 MDa [18–21]. Briefly, BN-PAGE is a discontinuous microscale electrophoretic technique to isolate microgram amounts of membrane protein complexes from biological membranes for use several uses, such as the clinical diagnostics of mitochondrial disorders, identification of protein–protein interactions, in-gel activity, protein import, etc. ([20], *and references within*).

For BN-PAGE, membranes are solubilized by nonionic (neutral), nondenaturing detergents, selected based on the detergent stability of the complexes of interest (e.g., digitonin is the mildest detergent, dodecyl maltoside has stronger delipidating properties, and Triton X-100 shows intermediate behavior). After solubilization, Coomassie Blue G-250 is added, which binds to the surface

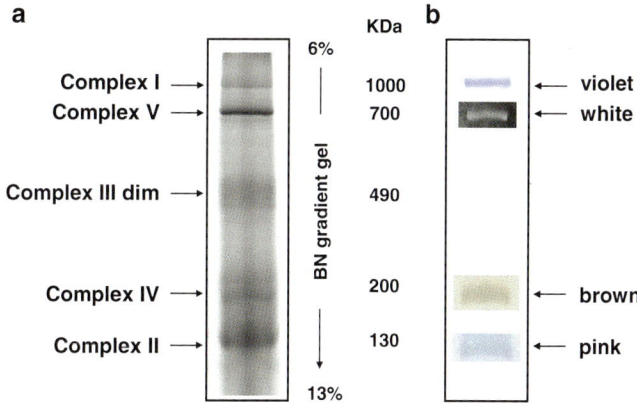

Fig. 1 Representative BN-PAGE separation of dodecylmaltoside-solubilized mitochondrial complexes. (**a**) Coomassie blue stained BN-PAGE 6–13 % gradient gel of crude rat liver mitochondria. Bands characteristic of individual OXPHOS complexes are recognizable. The native mass range is also reported. 15 μg of mitochondrial protein extract were loaded. (**b**) Histochemical staining of in-gel activity of individual OXPHOS complexes. The color of the specific band staining is reported

of the proteins and converts them into water-soluble molecules. This allows the negatively charged complexes to be separated according to molecular mass and detected as blue bands in the BN gels (in 1D BN gels, the five major Coomassie Blue-stained bands represent the oxidative phosphorylation complexes) (Fig. 1).

Supramolecular assemblies retained from the 1D BN-PAGE can be dissociated into individual complexes by applying an orthogonal modified BN-PAGE for a second native dimension, allowing the identification of interacting partners as well as their stoichiometric ratio [22]. Because 1D BN-PAGE separates intact OXPHOS complexes/supercomplexes, subsequent denaturing electrophoresis can resolve the individual subunits of the respective complexes. Indeed, OXPHOS assembly profiles can be obtained by two-dimensional blue native/SDS gel electrophoresis, which provides additional information on the role of specific proteins in the biogenesis of the OXPHOS system [20, 23].

Overall, BN-PAGE is a robust, manageable, reproducible, and cost- and time-efficient method, and it has the important advantage of being compatible with in-gel activity staining procedures specifically measuring the OXPHOS enzymes (whereas most spectrophotometric assays also detect other cellular activities) [24–27] (Fig. 1). A number of previous studies support how BN-PAGE, combined with histochemical staining, can provide valuable information for the clinical diagnosis of OXPHOS states by monitoring their catalytic and/or structural features [28–30]. Of note, the measurement of the specific activities in reactive bands has to be considered semiquantitative and strongly affected by the in-gel milieu [21, 31, 32]. The preparation of mitochondria from tissues such as liver, as well as the subsequent processing for BN-PAGE, can be problematic and actually differ somewhat from the protocols for muscle and brain tissue.

Here, we will describe the protocols used in our laboratory to isolate the mitochondria from mammalian liver and to study OXPHOS complexes by 1D BN-PAGE in terms of assembly and individual in-gel activity.

2 Materials

2.1 Materials and Equipment

1. Prepare all solutions using ultrapure [double distilled (dd)] water and analytical grade reagents (we did not add sodium azide to the reagents), and store all reagents at 4–8 °C (unless indicated otherwise). Diligently follow all waste disposal regulations.

2. The following pieces of equipment are required: a commercially available vertical electrophoresis apparatus [optional, (mini)gel multicasting chamber], a power supply (600 V, 500 mA), a motor-driven tightly fitting glass–Teflon homogenizer, a gradient mixer for use with a magnetic stirrer (optional peristaltic pump for casting acrylamide gradient gels), refrigerated centrifuges, and ultracentrifuges. We most often use minigel systems in our laboratory, which allow sufficiently good separation of the OXPHOS complexes and have the advantage of requiring fewer reagents and less sample material. This is important for downstream applications such as in-gel activity assays, which necessitate expensive chemicals.

2.2 Mitochondria Isolation and BN-PAGE Sample Preparation

1. Phenylmethylsulfonyl Fluoride (PMSF) Stock Solution: 0.5 M in Dimethylsulfoxide (DMSO).

2. Sucrose buffer: 440 mM sucrose, 20 mM MOPS, 1 mM EDTA, pH 7.2, 0.2 mM PMSF (to be added shortly before use).

3. NaCl solution: 500 mM NaCl, 10 mM Na^+/MOPS, pH 7.2.

4. Aminocaproic acid/Bis-Tris HCl solution: 1 M aminocaproic acid, 50 mM Bis-Tris HCl, pH 7.

5. Triton X-100 solution: 10 %.

6. Dodecyl maltoside solution: 10 % dodecyl maltoside (w/v) in water; store 1 ml aliquots at –20 °C.

7. Digitonin solution: 20 % digitonin (w/v) in water; store 0.1–1 ml aliquots at –20 °C. Heating (>70 °C) can be required for some lots of digitonin, but avoid repeated freezing/thawing, which can lead to the insolubility of digitonin at temperatures <50 °C.

8. Coomassie Blue G-250 solution: 5 % Coomassie Blue G-250 (w/v) in 1 M aminocaproic acid.

2.3 Gel Casting and Electrophoresis

1. Acrylamide solution (commercially available): 49.5 % (30 %) (w/v), acryl/bisacryl 32:1 (or 37.5:1). Acrylamide and bisacrylamide are highly neurotoxic. When handling these chemicals, wear gloves and use a pipetting aid.

2. Aminocaproic acid solution: 1 M aminocaproic acid, pH 7.

3. Bis-Tris HCl solution: 1 M Bis-Tris HCl, pH 7.

4. Ammonium persulphate (APS): 10 % APS, 1 g/10 ml; store aliquots at –20 °C and thaw fresh.

5. N,N,N',N'-tetramethylethylenediamine (TEMED): commercially available; use from stock bottle.

6. 5× cathode buffer blue: 250 mM tricine, 75 mM Bis-Tris, + 0.1 % (w/v) Coomassie Blue G-250, pH 7 (adjust at 4 °C).

7. 5× cathode buffer: 250 mM tricine, 75 mM Bis-Tris, pH 7 (adjust at 4 °C).

8. 6× anode buffer: 300 mM Bis-Tris, pH 7 (adjust at 4 °C).

9. NativeMark™ Unstained Protein Standard (Invitrogen).

2.4 Brilliant Blue G Gel Staining

1. Fixing solution: 7 % acetic acid, 40 % methanol in water.

2. Staining solution: Brilliant Blue G colloidal concentrate, according to the manufacturer instructions.

3. Destaining solution: 10 % acetic acid, 25 % methanol in water.

4. Rinsing solution: 25 % methanol in water.

2.5 Histochemical Evaluation of OXPHOS In-Gel Activities

1. Buffer for NADH dehydrogenase (ubiquinone) (*Complex I*) activity: 100 mM Tris–HCl, pH 7.4, 768 mM glycine, 0.1 mM β-NADH, 0.04 % nitrotetrazolium blue (NTB) (w/v).

2. Buffer for succinate dehydrogenase (*Complex II*) activity: 100 mM Tris–HCl, pH 7.4, 100 mM glycine, 10 mM succinate, 0.1 % NTB (w/v).

3. Buffer for cytochrome c oxidase (*Complex IV*) activity (for 10 ml): 5 mg 3,3′-Diaminobenzidine tetrahydrochloride (DAB) in 9 ml phosphate buffer (0.05 M, pH 7.4), 1 ml 20 µg/ml catalase, 10 mg cytochrome c, 750 mg sucrose.

4. Buffer for FoF1ATPase (*Complex V*) activity: 35 mM Tris–HCl, pH 7.8, 270 mM glycine, 14 mM $MgSO_4$, 0.2 % b$(NO_3)_2$, 8 mM ATP.

5. Fixing solution for BN gels: 50 % methanol, 10 % acetic acid in water.

6. Stop solution: 50 % methanol in water.

3   Methods

Carry out all procedures at 4 °C unless otherwise specified.

3.1   Isolation of Mitochondria

1. Mince 50 mg wet weight liver tissue in ice-cold sucrose buffer, and rinse several times until the solution is clear.

2. Gently homogenize the minced liver samples for 1 min in 0.5 ml sucrose buffer by using a tightly fitting glass–Teflon homogenizer (set at 500 rpm).

3. Pass the homogenate through a sterile gauze and adjust the volume [a dilution of 1:10 (w/v) is recommended].

4. Isolate the nuclear pellet by centrifuging at $500 \times g$ for 10 min.

5. Remove the supernatant and place in a new labeled and pre-cooled tube.

6. Centrifuge the supernatant again at $3,000 \times g$ for 10 min. Remove the supernatant again and reserve the pellet. This contains the cytosolic and membrane fractions that also contain lighter mitochondria (*see* **Notes 1** and **2**).

7. Wash the pellet from **step 6** twice in ice-cold sucrose buffer, and resuspend it by pipetting.

8. At this time point, a protein concentration assay is recommended (*see* **Note 3**).

3.2   Solubilization of OXPHOS Respiratory Complexes from Crude Mitochondrial Pellet

1. Homogenize the mitochondrial sediment in 0.5 ml NaCl solution.

2. Centrifuge at $20,000 \times g$ for 20 min; discard the supernatant.

3. Add 0.150 ml aminocaproic acid/Bis-Tris HCl solution, 0.020 ml 10 % Triton X-100, and homogenize by twirling with a tiny spatula.

4. Centrifuge at $100,000 \times g$ for 15 min; discard the supernatant.

5. Add 0.04 ml aminocaproic acid/Bis-Tris HCl solution and homogenize.

6. Add 0.020 ml dodecyl maltoside solution. Optional: add 0.020 ml digitonin solution.

7. Add 0.010 ml Coomassie Blue G-250 solution (at this time, samples can be stored at –80 °C for several weeks) (*see* **Note 4**).

8. Apply to gel wells (volumes to be applied depend on gel and comb dimensions; 5–20 μl should be loaded into minigel wells).

3.3   Casting of Acrylamide Gradient Gels

1. Select and cast the appropriate gel type for BN-PAGE, noting that acrylamide gradient gels are commonly used for this procedure. First, cast separating gels at 4–8 °C using a gradient mixer. The amounts of each reagent needed to make ten

Table 1
Recipe for the casting of ten 6–13 % polyacrylamide gradient minigels (using the BioRad multicasting chamber, 1 mm)

6 % acrylamide	13 % acrylamide
7.6 ml acrylamide solution	14 ml acrylamide solution
9 ml dd water	0.2 ml dd water
19 ml aminocaproic acid solution (1 M, pH 7)	16 ml aminocaproic acid solution (1 M, pH 7)
1.9 ml Bis-Tris solution (1 M, pH 7)	1.6 ml Bis-Tris solution (1 M, pH 7)
200 μl 10 % APS	200 μl 10 % APS
20 μl TEMED	20 μl TEMED
Total volume: 38 ml	Total volume: 32 ml

Table 2
Recipe for the casting of ten 4 % polyacrylamide stacking gels (for mini gels)

4 % Stacking gel
3.5 ml acrylamide solution
8 ml dd water
12.5 ml aminocaproic acid solution (1 M, pH 7)
1.5 ml Bis-Tris solution (1 M, pH 7)
200 μl 10 % APS
50 μl TEMED
Total volume: 25.5 ml

6–13 % polyacrylamide gradient gels ($10 \times 100 \times 1.0$ mm) are given in Table 1 (*see* **Note 5**).

2. After casting the separating gel, cover it with water or isopropanol and allow it to polymerize at room temperature, which takes approximately 30–60 min.

3. When the gel has polymerized, remove the water or isopropanol (in this case, rinse with water) from the top of the gradient gels, remove from casting chamber, and cast the stacking (or sample) gel at room temperature (Table 2). Add the appropriate sample combs.

4. Allow the gel to polymerize for approximately 15–30 min at room temperature.

5. Remove the sample gel comb and overlay the gel with cathode buffer blue (1×) (*see* **Note 6**).

3.4 Electrophoresis	BN-PAGE using cathode buffer blue (i.e., with 0.02 % Coomassie Blue G-250 added) is performed at 4 °C in a vertical apparatus. After the samples are applied to the gel, the gel is run at 80 V until the samples have completely entered the sample gel (approximately 30 min for minigels). The gel is then run for several hours at 100–150 V (for minigels) until the blue dye has almost run off the bottom of the gel (*see* **Note 7**).

3.5 Gel Staining with Brilliant Blue G

To visualize and quantify the obtained electrophoretic profile and to intensify faint bands, BN gels can be stained with Brilliant Blue G.

1. After the runs are complete, immerse the gels in fixing solution (10–20 ml of solution is sufficient for minigels) and incubate at room temperature with gentle agitation for at least 30 min.

2. Incubate the gels in Brilliant Blue G according to the manufacturer instructions.

3. Incubate the gel in destaining solution with gentle agitation for 30–60 s, depending on gel dimensions (30 s is sufficient for minigels) at room temperature (*see* **Note 8**). Rinse the gels by incubating them in rinsing solution with gentle agitation at room temperature for 15 min. Repeat this step four times.

4. Wash twice with dd water at room temperature for 5 min per wash.

5. Scan the gels using a commercially available imaging system.

3.6 Quantification of Native OXPHOS Protein In-Gel Activity by Histochemical Evaluation

After the runs are complete, the following steps are initiated.

1. To visualize Complex I in-gel activity, incubate the gel slices in Complex I buffer with gentle agitation at room temperature. The required volume will depend on the gel-slice dimensions (10–20 ml is sufficient for minigels). The band color (violet) needs few minutes to appear and be detected. Preserve the original color of the reacting bands by fixing the gel slices in BN gel fixing solution (with gentle agitation at room temperature for 30 min) (*see* **Notes 9** and **10**).

2. To visualize Complex II in-gel activity, incubate the gel slices in Complex II buffer with gentle agitation at room temperature. The required volume will depend on the gel-slice dimensions. The band color (pink) needs 10–30 min to appear and be detected. Preserve the original color of the reacting bands by fixing the gel slices in BN gel fixing solution (with gentle agitation at room temperature for 30 min) (*see* **Notes 9–11**).

3. To visualize Complex IV in-gel activity, incubate gel slices with Complex IV buffer with gentle agitation at room temperature. The band color (brown) needs 15–60 min to appear and be detected. Preserve the original color of the reacting bands by fixing the gel slices in BN gel fixing solution (with gentle agitation at room temperature for 30 min) (*see* **Notes 9** and **10**).

4. Determine Complex V activity by incubation of BN-PAGE gels in Complex V buffer with gentle agitation at room temperature. The band color (white) needs several hours to appear and be detected. Stop the reaction in 50 % methanol, and wash the gels with distilled water.

5. Scan the gels using an imaging system, and express the areas of the colored bands as absolute values (arbitrary units).

4 Notes

1. When working on subcellular compartments, as in the case of mitochondria, the purity of the preparation is critical. Indeed, during mitochondrial preparation, several proteins can be co-isolated (i.e., from endoplasmic reticulum or other organelle membranes in close contact with the mitochondria), impairing quantitative and qualitative analyzes and thus altering data interpretation. Several methods have been applied in an effort to obtain pure mitochondria from tissues and cells, including differential centrifugation; density gradient centrifugation with Percoll™ [33], Nycodenz [34], Metrizamide [35], or sucrose [36] free-flow electrophoresis [37]; and kit-based methods [38]. Mitochondria can also be highly purified by immunoisolation (by means of mitochondria-specific antibodies), although the costs are high and sample retrieval is low [39]. The purity of the mitochondrial preparation—a fundamental issue above all others when performing quantitative analyzes of proteins with different subcellular localizations—can be tested by measuring marker-enzyme activities (such as citrate synthase) and/or by western blotting for specific markers [i.e., voltage-dependent anion channels (VDACs) as positive control or histones as negative control].

2. Mitochondrial heterogeneity with respect to sedimentation characteristics, chemical makeup, and enzyme activities has long been recognized. An optional method to isolate mitochondria to distinguish "heavy" ($1,000 \times g$), "medium" ($3,000 \times g$), and "light" ($10,000 \times g$) mitochondrial subfractions has been suggested [40]. The $3,000 \times g$ fraction represents relatively intact mitochondria and is characterized by high oxidation rates and good respiratory control. Even lighter liver mitochondria can be isolated at $27,000 \times g$. However, in this case, the high gravitational field can significantly affect the integrity of the obtained fraction [41].

To obtain the light fraction, the supernatant from the centrifugation at $3,000 \times g$ has to be centrifuged at $10,000 \times g$ for 10 min at 4 °C. To obtain the lighter fraction, the supernatant obtained at $10,000 \times g$ has to be centrifuged at $27,000 \times g$ for

10 min at 4 °C. The pellets obtained in these two separate steps have to be washed twice by adding the sucrose buffer again and then resuspending by pipetting. Then, the suspension is centrifuged again for 10 min either at 10,000 or at 27,000 × g. Resuspend the mitochondria pellets in sucrose buffer.

3. At this step, the protocol can be temporarily stopped by freezing the mitochondrial preparation before membrane solubilization. A freeze–thaw cycle helps to break the membrane and facilitates respiratory chain complexes solubilization during the subsequent steps for BN-PAGE.

4. The samples can be stored at –80 °C for several months before loading on the gel.

5. Commercial ready-to-use blue native gels are also available (Invitrogen, Native PAGE Novex Bis-Tris Gel system).

6. Gels can be stored at 4 °C for 1 month if the overlay buffer is renewed at ~3-day intervals.

7. For better band detection, the cathode buffer blue can be removed after one-third of the gel is run and replaced with cathode buffer before completing the electrophoresis.

8. Destaining is important for reducing the gel background signal and the signal from nonspecific staining.

9. The time latency for color appearance for tissues other than liver can be longer. In any case, in vivo iodothyronine administration or changes in the thyroid state of the whole animal can influence time latency for color appearance.

10. To save materials and reagents, the gel can be cut to the band of interest.

11. The Coomassie Blue G-250 dye, used in BN-PAGE, interferes with in-gel fluorescence detection and some in-gel catalytic activity assays, as in the case of Complex III. This problem can be overcome by performing high-resolution clear native electrophoresis (CNE). The basic separation principles and the performance of BN-PAGE and CNE are markedly different, although the experimental setups are largely identical except for the addition of Coomassie Blue G-250 dye to the sample and cathode buffer for BN-PAGE. CNE, which omits Coomassie Blue G-250, lacks the advantages of the charge shift technique BN-PAGE, is limited to the separation of acidic proteins with a pI below the pH of the gels, and is often characterized by protein aggregation during electrophoresis and by a significantly lower resolution compared to BN-PAGE. However, it offers clear advantages for in-gel catalytic activity assays and detection of fluorescent labels. The first in-gel histochemical staining protocol for respiratory Complex III has been reported [42].

References

1. Seo AY, Joseph AM, Dutta D et al (2010) New insights into the role of mitochondria in aging: mitochondrial dynamics and more. J Cell Sci 123:2533–2542

2. Chen H, Chan DC (2010) Physiological functions of mitochondrial fusion. Ann N Y Acad Sci 1201:21–25

3. Goglia F, Moreno M, Lanni A (1999) Action of thyroid hormones at the cellular level: the mitochondrial target. FEBS Lett 452:115–120

4. Cioffi F, Lanni A, Goglia F (2010) Thyroid hormones, mitochondrial bioenergetics and lipid handling. Curr Opin Endocrinol Diabetes Obes 17:402–407

5. Weitzel JM, Iwen KA (2011) Coordination of mitochondrial biogenesis by thyroid hormone. Mol Cell Endocrinol 342:1–7

6. Lanni A, Moreno M, Lombardi A et al (2005) 3,5-diiodo-L-thyronine powerfully reduces adiposity in rats by increasing the burning of fats. FASEB J 19:1552–1554

7. Silvestri E, Cioffi F, Glinni D et al (2010) Pathways affected by 3,5-diiodo-l-thyronine in liver of high fat-fed rats: evidence from two-dimensional electrophoresis, blue-native PAGE, and mass spectrometry. Mol Biosyst 6:2256–2271

8. de Lange P, Cioffi F, Senese R et al (2011) Nonthyrotoxic prevention of diet-induced insulin resistance by 3,5-diiodo-L-thyronine in rats. Diabetes 60:2730–2739

9. Perra A, Simbula G, Simbula M et al (2008) Thyroid hormone (T3) and TRbeta agonist GC-1 inhibit/reverse nonalcoholic fatty liver in rats. FASEB J 22:2981–2989

10. Mollica MP, Lionetti L, Moreno M et al (2009) 3,5-diiodo-l-thyronine, by modulating mitochondrial functions, reverses hepatic fat accumulation in rats fed a high-fat diet. J Hepatol 51:363–370

11. Silvestri E, Lombardi A, Glinni D et al (2011) Mammalian mitochondrial proteome and its functions: current investigative techniques and future perspectives on ageing and diabetes. J Integr OMICS. doi:10.5584/jiomics.v1i1.51

12. Da Cruz S, Parone PA, Martinou JC (2005) Building the mitochondrial proteome. Expert Rev Proteomics 2:541–551

13. Silvestri E, Moreno M, Schiavo L et al (2006) A proteomics approach to identify protein expression changes in rat liver following administration of 3,5,3'-triiodo-L-thyronine. J Proteome Res 5:2317–2327

14. Silvestri E, Burrone L, de Lange P et al (2007) Thyroid-state influence on protein-expression profile of rat skeletal muscle. J Proteome Res 6:3187–3196

15. Moreno M, Silvestri E, De Matteis R et al (2011) 3,5-Diiodo-L-thyronine prevents high-fat-diet-induced insulin resistance in rat skeletal muscle through metabolic and structural adaptations. FASEB J 25:3312–3324

16. Lombardi A, Silvestri E, Cioffi F et al (2009) Defining the transcriptomic and proteomic profiles of rat ageing skeletal muscle by the use of a cDNA array, 2D- and Blue native-PAGE approach. J Proteomics 72:708–721

17. Silvestri E, Glinni D, Cioffi F et al (2012) Metabolic effects of the iodothyronine functional analogue TRC150094 on the liver and skeletal muscle of high-fat diet fed overweight rats: an integrated proteomic study. Mol Biosyst 8:1987–2000

18. Schägger H, von Jagow G (1991) Blue native electrophoresis for isolation of membrane protein complexes in enzymatically active form. Anal Biochem 199:223–231

19. Schagger H, Cramer WA, Jagow G (1994) Analysis of molecular masses and oligomeric states of protein complexes by blue native electrophoresis and isolation of membrane protein complexes by two-dimensional native electrophoresis. Anal Biochem 217:220–230

20. Wittig I, Braun HP, Schägger H (2006) Blue native PAGE. Nat Protoc 1:418–428

21. Reisinger V, Eichacker LA (2008) Isolation of membrane protein complexes by blue native electrophoresis. Methods Mol Biol 424:423–431

22. Schägger H, Pfeiffer K (2000) Supercomplexes in the respiratory chains of yeast and mammalian mitochondria. EMBO J 19:1777–1783

23. Calvaruso MA, Smeitink J, Nijtmans L (2008) Electrophoresis techniques to investigate defects in oxidative phosphorylation. Methods 46:281–287

24. Zerbetto E, Vergani L, Dabbeni-Sala F (1997) Quantification of muscle mitochondrial oxidative phosphorylation enzymes via histochemical staining of blue native polyacrylamide gels. Electrophoresis 18:2059–2064

25. Jung C, Higgins CM, Xu Z (2000) Measuring the quantity and activity of mitochondrial electron transport chain complexes in tissues of central nervous system using blue native polyacrylamide gel electrophoresis. Anal Biochem 286:214–223

26. Eubel H, Heinemeyer J, Sunderhaus S et al (2004) Respiratory chain supercomplexes in plant mitochondria. Plant Physiol Biochem 42:937–942

27. Reifschneider NH, Goto S, Nakamoto H et al (2006) Defining the mitochondrial proteomes from five rat organs in a physiologically significant context using 2D blue-native/SDS-PAGE. J Proteome Res 5:1117–1132

28. Krause F, Reifschneider NH, Goto S et al (2005) Active oligomeric ATP synthases in mammalian mitochondria. Biochem Biophys Res Commun 329:583–590

29. Bisetto E, Comelli M, Salzano AM et al (2013) Proteomic analysis of F1F0-ATP synthase super-assembly in mitochondria of cardiomyoblasts undergoing differentiation to the cardiac lineage. Biochim Biophys Acta 1827:807–816

30. Wüst RC, Myers DS, Stones R et al (2012) Regional skeletal muscle remodeling and mitochondrial dysfunction in right ventricular heart failure. Am J Physiol Heart Circ Physiol 302:H402–H411

31. Krause F (2006) Detection and analysis of protein-protein interactions in organellar and prokaryotic proteomes by native gel electrophoresis: (Membrane) protein complexes and supercomplexes. Electrophoresis 27:2759–2781

32. Wittig I, Schägger H (2008) Features and applications of blue-native and clear-native electrophoresis. Proteomics 8:3974–3990

33. Pagliarini DJ, Calvo SE, Chang B et al (2008) A mitochondrial protein compendium elucidates complex I disease biology. Cell 134:112–123

34. Li J, Cai T, Wu P et al (2009) Proteomic analysis of mitochondria from *Caenorhabditis elegans*. Proteomics 9:4539–4553

35. Taylor SW, Fahy E, Zhang B et al (2003) Characterization of the human heart mitochondrial proteome. Nat Biotechnol 21: 281–286

36. Rezaul K, Wu L, Mayya V et al (2005) A systematic characterization of mitochondrial proteome from human T leukemia cells. Cell Proteomics 4:169–181

37. Zischka H, Weber G, Weber PJ et al (2003) Improved proteome analysis of *Saccharomyces cerevisiae* mitochondria by free-flow electrophoresis. Proteomics 3:906–916

38. Hartwig S, Feckler C, Lehr S et al (2009) A critical comparison between two classical and a kit-based method for mitochondria isolation. Proteomics 9:3209–3214

39. Herrnstadt C, Clevenger W, Ghosh SS et al (1999) A novel mitochondrial DNA-like sequence in the human nuclear genome. Genomics 60:67–77

40. Lanni A, Moreno M, Lombardi A et al (1996) Biochemical and functional differences in rat liver mitochondrial subpopulations obtained at different gravitational forces. Int J Biochem Cell Biol 28:337–343

41. Goglia F, Liverini G, Lanni A et al (1988) Light mitochondria and cellular thermogenesis. Biochem Biophys Res Commun 151: 1241–1249

42. Wittig I, Karas M, Schägger H (2007) High resolution clear native electrophoresis for in-gel functional assays and fluorescence studies of membrane protein complexes. Mol Cell Proteomics 6:1215–1225

Chapter 11

Detection of UCP1 Protein and Measurements of Dependent GDP-Sensitive Proton Leak in Non-phosphorylating Thymus Mitochondria

Kieran J. Clarke, Audrey M. Carroll, Gemma O'Brien, and Richard K. Porter

Abstract

Over several years we have provided evidence that uncoupling protein 1 (UCP1) is present in thymus mitochondria. We have demonstrated the conclusive evidence for the presence of UCP1 in thymus mitochondria and we have been able to demonstrate a GDP-sensitive UCP1-dependent proton leak in non-phosphorylating thymus mitochondria. In this chapter, we show how to detect UCP1 in mitochondria isolated from whole thymus using immunoblotting. We show how to measure GDP-sensitive UCP1-dependent oxygen consumption in non-phosphorylating thymus mitochondria and we show that increased reactive oxygen species production occurs on addition of GDP to non-phosphorylating thymus mitochondria. We conclude that reactive oxygen species production rate can be used as a surrogate for detecting UCP1 catalyzed proton leak activity in thymus mitochondria.

Key words Thymus mitochondria, UCP1, GDP-sensitive, Proton leak, Reactive oxygen species, Amplex Red, Oxygen consumption, Immunoblotting, Peptide polyclonal antibodies

---

1 Introduction

Uncoupling protein 1 (UCP1) was thought to be a unique to BAT [1]. However, recently in our laboratory we have demonstrated that UCP1 is present in the thymus. Data supporting the presence of UCP1 in thymus mitochondria includes (1) evidence of transcript for UCP1 in thymus, (2) immunoblot detection of UCP1 in thymus mitochondria from wild type but not UCP1 knockout mice using peptide polyclonal antibodies to UCP1, (3) purification (and MALDI identification) of UCP1 from thymus, (4) GDP-binding studies to thymus mitochondria, (5) GDP-sensitive oxygen consumption by non-phosphorylating thymus mitochondria by wild type but not UCP1 knockout mice, (6) increased reactive oxygen species production in thymus mitochondria from UCP1 knockout mice compared

Carlos M. Palmeira and Anabela P. Rolo (eds.), *Mitochondrial Regulation*, Methods in Molecular Biology, vol. 1241, DOI 10.1007/978-1-4939-1875-1_11, © Springer Science+Business Media New York 2015

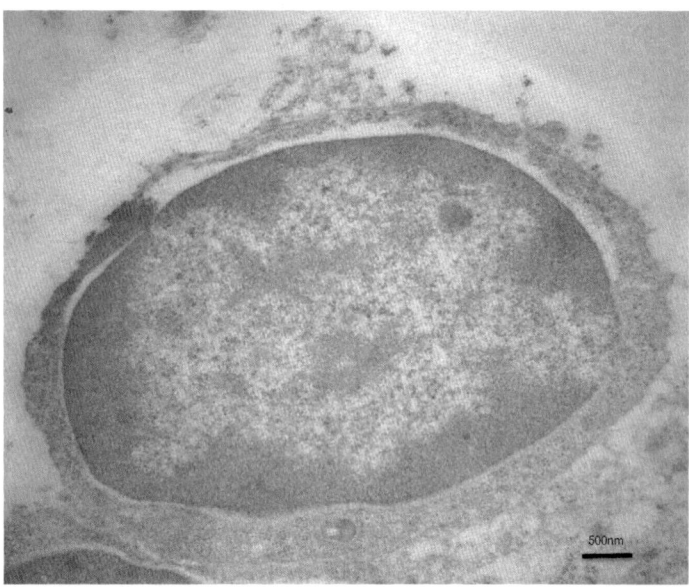

Fig. 1 Transmission electron micrograph of a rat thymocyte

to wild-type controls, and (7) confocal images of UCP1 associated with mitochondrial in thymocytes from wild type but not UCP1 knockout mice [2–8]. It should also be noted that thymocytes are very small (5 μm in diameter) (Fig. 1) and thus are easily distinguishable from brown adipocytes (50–70 μm in diameter). Mitochondrial proton leak rate in all mitochondria can be measured, indirectly, using the oxygen consumption of non-phosphorylating mitochondria. We have been able to demonstrate a GDP-sensitive proton leak in non-phosphorylating thymus mitochondria due to UCP1. Inhibiting UCP1 would be predicted to increase reactive oxygen species production from the electron transport chain in non-phosphorylating mitochondria due to the resultant increase in membrane potential. We showed that increased reactive oxygen species production does occur on addition of GDP to non-phosphorylating thymus mitochondria and thus increased reactive oxygen species production can be used as a surrogate index for detecting UCP1 activity in thymus mitochondria. We also demonstrate how to detect UCP1 in mitochondria isolated from whole thymus by an immunoblot.

2 Materials

All solutions were prepared using ultrapure water from a Millipore Elix Advantage 10 Water Purification System (resistivity >5 MΩ cm) and analytical grade reagents. All reagents were stored at room temperature (unless indicated otherwise). We adhered to waste disposal regulations rigorously.

2.1 Preparative Steps for Isolation of Thymus Mitochondria

1. Isolation Buffer: At least a day before isolation of mitochondria, weigh out 85.75 g sucrose, 121.14 g Tris and 0.38 g EGTA. Add 900 ml water. Mix and adjust the pH to 7.4 with concentrated HCl. Make the isolation medium up to 1 L with water and store at 4 °C (*see* **Note 1**).

2. Isolation apparatus: Place a Dounce/Potter Homogeniser and pestles (covered in tin foil) on ice (*see* **Note 2**).

3. Thymus: Ten young male/female Wistar rats c.12 weeks old/190 g (*see* **Note 3**).

2.2 Components for Sodium Dodecyl Sulfate Polyacrylamide Gel Electrophoresis (SDS-PAGE)

1. The gel apparatus used was a Mini-PROTEAN® 3 CELL (Bio-Rad).

2. Resolving gel buffer: 1.5 M Tris–HCl, pH 8.8 (Add ~100 ml of water to a 1 l beaker). Weight 181.7 g Tris and transfer to the beaker. Add water to a volume of 900 ml. Mix and adjust pH with HCL. Make up to 1 l with water. Store at 4 °C.

3. Stacking gel buffer: 0.5 M Tris–HCl, pH 6.8. Weigh 60.6 g Tris and prepare a 1 L solution as in previous step. Store at 4 °C.

4. Protogel™ ultra-pure 30 % (w/v) acrylamide and 0.8 % (v/v) bisacrylamide solution (National Diagnostics).

5. 10 % (w/v) ammonium persulfate (see **Note 4**).

6. N, N, N, N'-tetramethyl-ethylenediamine (TEMED) (Sigma). Store at 4 °C.

7. A glycine-based running buffer (0.38 M glycine, 0.05 M Tris, 0.1 % (w/v) SDS) was used for electrophoresis.

8. The sample buffer was made up as 4× strength. Components of the sample buffer are as follows: water (3.8 ml), 0.5 M Tris–HCl, pH 6.8 (1.0 ml), glycerol (0.8 ml), 10 % (w/v) SDS (1.6 ml) and 1 % (w/v) Bromophenol blue (0.4 ml). 5 % β-mercaptoethanol (v/v) was added to the sample buffer immediately prior to use (see **Note 5**).

9. Bromophenol blue solution: Dissolve 0.1 g BPB in 10 mL water for a 1 % (w/v) stock solution.

2.3 Components for Immunoblotting

1. Polyvinylidene difluoride (PVDF) membranes (Immobilon-PSQ; Millipore) (see **Note 6**).

2. Soak six pieces of blotting paper (cut to the size of the gel) in transfer buffer for 5 min.

3. Immunoblot transfer buffer: 0.192 M glycine, 0.025 M Tris–HCl, pH 8.3, 0.013 M SDS, 15 % (v/v) methanol.

4. Phosphate Buffered Saline (PBS): 0.14 M NaCl, 2.7 mM KCl, 11.5 mM Na_2PO_4, 1.8 mM KH_2PO_4, pH 7.4.

5. Blocking/Diluent solution: PBS containing 0.1 % Tween20 and 5 % (w/v) Marvel milk powder (*see* **Note 7**). Store at 4 °C.

6. Transfer was achieved using a semidry transfer apparatus (Hoeffer) at 110 mA for 2 h (*see* **Note 8**).

7. Ponceau S solution (0.25 % (w/v) Ponceau S, in 3 % (v/v) trichloroacetic acid) (*see* **Note 9**).

2.4 Molecular Weight Standards and Antibodies

1. A broad range set of prestained molecular weight marker standards (provided by New England Biolabs) were used. The prestained markers used were *E. coli* MBP-β-galactosidase (175 kDa), *E. coli* MBP-paramyosin (83 kDa), bovine liver glutamic dehydrogenase (62 kDa), rabbit muscle aldolase (47.5 kDa), rabbit muscle triosephosphate isomerase (32.5 kDa), bovine milk β-lactoglobulin A (25 kDa), chicken egg white lysozyme (16.5 kDa), and bovine lung aprotinin (6.5 kDa).

2. Rabbit anti-UCP1 antibody: we used an antibody to an antigenic peptide region (145–158) of rat UCP1 was used (*see* **Note 10**).

3. Rabbit anti-ATP synthase antibody: we used an antibody to the β-subunit of the F_1-ATP synthase (*see* **Note 11**).

4. Secondary antibodies were horse-radish peroxidase (HRP) conjugated goat anti-rabbit secondary antibody.

2.5 Incubation Media and Apparati for Measurement of Oxygen Consumption and Reactive Oxygen Species Production by Thymus Mitochondria

1. The basic incubation medium for thymus mitochondria is predominantly an ionic medium containing 120 mM KCl, 5 mM Hepes, 1 mM EGTA and 0.1 % (w/v) defatted bovine serum albumin, pH 7.4 (with KOH).

2. The Oroboros Respirometer was used to measure oxygen consumption by thymus mitochondria (*see* **Note 12**).

3. Stocks of the following solutions are required to give final concentrations of 1 µg/ml oligomycin, 5 µM atractyloside and 10 mM glycerol-3-phosphate in the incubation medium for oxygen consumption rate measurements by non-phosphorylating thymus mitochondria.

4. Amplex Red conversion to Resorufin was used to detect reactive oxygen species production by mitochondria (*see* **Note 13**).

5. Stocks of the following solutions are required to final concentrations of 1 µg/ml oligomycin, 5 µM atractyloside, 10 mM glycerol-3-phosphate, 5 µM Amplex Red, 10 U/ml horse-radish peroxidase, and 30 U/ml superoxide dismutase for measurements of reactive oxygen species production by non-phosphorylating thymus mitochondria.

6. Stocks of the protonophore carbonyl cyanide *p*-(trifluoro-methoxy) phenyl-hydrazone (FCCP) and the UCP1 inhibitor GDP are required for addition to thymus mitochondria to give final concentrations of 0.5 µM and 1 mM respectively.

3 Methods

3.1 Isolation of Thymus Mitochondria

1. Rats were euthanized by carbon dioxide asphyxiation (*see* **Note 14**) in accordance with strict guidelines on animal welfare as outlined in Directive 2010/63/EU and S.I No. 543 2012.

2. The thymus was removed from the abdominal cavity, trimmed of connective tissue and fat, placed into a beaker containing ice cold (0–4 °C) isolation buffer. Any brown adipose tissue (BAT) (*see* **Note 15**) present in the vicinity of the thymus is clearly visible and distinguishable from the thymus (Fig. 2) and is removed prior to the thymus being weighed.

3. Thymus mitochondria were prepared essentially by the method of Chappell [9] (*see* **Note 16**).

4. The thymus was chopped finely using a scissors and washed several times using ice-cold isolation buffer.

5. The tissue was poured into a glass homogeniser tube to a final volume of approximately 40 ml. The tissue was then homogenized by hand with four passes using a pestle of 0.26 in. (loose) clearance followed by six passes using a pestle of 0.12 in. (tight) clearance.

6. The homogenate was filtered through four layers of muslin.

7. The resulting filtrate was centrifuged at $800 \times g$ for 3 min at 4 °C. The pellet was discarded and the supernatant (containing organellar and cytosolic fractions) was centrifuged at $12,000 \times g$ for 10 min at 4 °C.

8. The supernatant was discarded and the pellet (containing an enriched mitochondrial fraction) was re-suspended in isolation buffer containing 2 % (w/v) de-fatted BSA (*see* **Note 17**) and centrifuged at $12,000 \times g$ for 10 min at 4 °C.

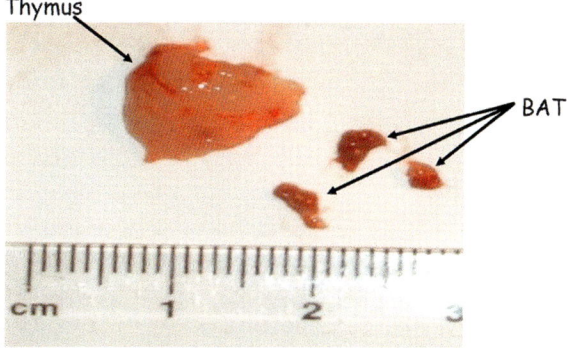

Fig. 2 A thymus and associated BAT excised from a rat

9. This pellet was subsequently resuspended in isolation buffer (without BSA) and centrifuged at $12,000 \times g$ for a further 10 min at 4 °C (*see* **Note 18**).

10. The resulting mitochondrial pellet was resuspended in 0.05 ml isolation buffer per gram of original tissue and mitochondria used within 6–8 h of isolation.

11. A protein assay was used to determine the concentration of mitochondrial protein in the final suspension (*see* **Note 19**).

3.2 SDS PAGE Electrophoresis

1. Sodium dodecyl sulfate-polyacrylamide gel electrophoresis (SDS-PAGE) was performed by the method of Laemmli [13], using a Mini-PROTEAN® 3 CELL (Bio-Rad) with a 5 % stacking and a 12 % resolving gel.

2. The 12 % resolving gel was composed as follows: water (3.35 ml), 1.5 M Tris–HCl, pH 8.8 (2.5 ml), 10 % (w/v) SDS (100 µl), Protogel™ (ultra-pure 30 % (w/v) acrylamide and 0.8 % (v/v) bisacrylamide solution; National Diagnostics; 4.0 ml), TEMED (5 µl) and 10 % (w/v) ammonium persulfate (50 µl). Cast the gel (*see* **Note 20**).

3. The stacking gel (5 %) was prepared with water (6.1 ml), 0.5 M Tris–HCl, pH 6.8 (2.5 ml), 10 % (w/v) SDS (100 µl), Protogel™ (ultra-pure 30 % (w/v) acrylamide, and 0.8 % (v/v) bisacrylamide solution; National Diagnostics; 1.33 ml), TEMED (10 µl) and 10 % (w/v) ammonium persulfate (50 µl). Insert a 10-well gel comb.

4. The sample buffer was made up as 4× strength. Components of the sample buffer are as follows: water (3.8 ml), 0.5 M Tris–HCl, pH 6.8 (1.0 ml), glycerol (0.8 ml), 10 % (w/v) SDS (1.6 ml) and 1 % (w/v) Bromophenol blue (0.4 ml). 5 % β-mercaptoethanol (v/v) was added to the sample buffer immediately prior to use.

5. The sample buffer was added to the mitochondrial protein so that the final concentration of sample buffer was 1×. Once the sample buffer and mitochondrial protein were added together, the samples were vortexed briefly and boiled for 5 min on a heating block set to 100 °C. The samples were then pulsed in a bench top centrifuge, loaded into separate wells.

6. A glycine-based running buffer (0.38 M glycine, 0.05 M Tris, 0.1 % (w/v) SDS) was used for electrophoresis. Electrophorese was at a constant current (200 V) for 45 min, until the tracker dye reached the bottom of the resolving gel. Gels were then prepared for immunoblotting.

3.3 Immunoblotting

1. Following SDS-PAGE, resolved proteins were transferred onto polyvinylidene difluoride (PVDF) membranes (Immobilon-P^SQ; Millipore).

2. Wet the PVDF membrane in 100 % methanol for ~5–10 s.

3. Drain and equilibrate the PVDF membrane and pieces of filter paper in transfer buffer for 5 min.

4. Transfer was achieved using a semidry transfer apparatus (Hoeffer) at 110 mA for 2 h. Firstly, the stacking gel was removed and gels were rinsed briefly in semidry transfer buffer (0.192 M glycine, 0.025 M Tris–HCl, pH 8.3, 0.013 M SDS, 15 % (v/v) methanol) and carefully arranged in the semidry transfer apparatus as directed by the manufacturers.

5. After transfer was complete, blotted proteins on the PVDF membrane were directly incubated in Ponceau S solution (0.25 % (w/v) Ponceau S, in 3 % (v/v) trichloroacetic acid) and washed gently with distilled water.

6. The blot was then washed in phosphate-buffered saline (PBS) (0.14 M NaCl, 2.7 mM KCl, 11.5 mM Na_2HPO_4, 1.8 mM KH_2PO_4, pH 7.4).

7. Blocking of the membrane was performed by incubating the blot in PWB (PBS containing 0.1 % (w/v) Tween 20) containing 5 % (w/v) Marvel milk powder at room temperature for 1 h or overnight at 4 °C. This blocking was followed by 3×10 min washes using PWB.

8. Blots were then incubated in primary antibody (in PWB containing 5 % (w/v) Marvel milk powder) overnight at 4 °C or at room temperature for 1 h containing a 1:1,000 dilution of an affinity-purified Eurogentec anti-UCP 1 peptide antibody (or a 1:1,000 dilution or a commercial anti-UCP 1 peptide antibody (Sigma/Calbiochem)). Well loading/mitochondrial protein loading was determined using the antibody to the β-subunit of the F_1-ATP synthase ($F_1β$) (1:1,000 dilution).

9. Following this primary antibody incubation, the blots were washed for 3×10 min in PWB.

10. The blots were then incubated with a horse-radish peroxidase (HRP) conjugated goat anti-rabbit secondary antibody (1:10,000 dilution) in PWB containing 5 % Marvel milk powder for 1 h at room temperature.

11. Following this, blots were further washed for 3×10 min in PWB. Blots were developed using an enhanced chemiluminescence (ECL) detection system (Amersham-Pharmacia) for detecting horse-radish peroxidase labelled antibody, by means of the HRP catalyzed oxidation of luminol under alkaline conditions.

12. Results were visualized by exposure to Kodak X-Omat LS film (Figs. 3 and 4).

13. The quality of the antibody was demonstrated by the observation that increased amounts of UCP1 were detected with increased concentrations of BAT (but not liver) liver mitochondria from C57BL/6J wild type as expected. No UCP 1 was detected in BAT mitochondria from UCP1 knockout mice (Fig. 3).

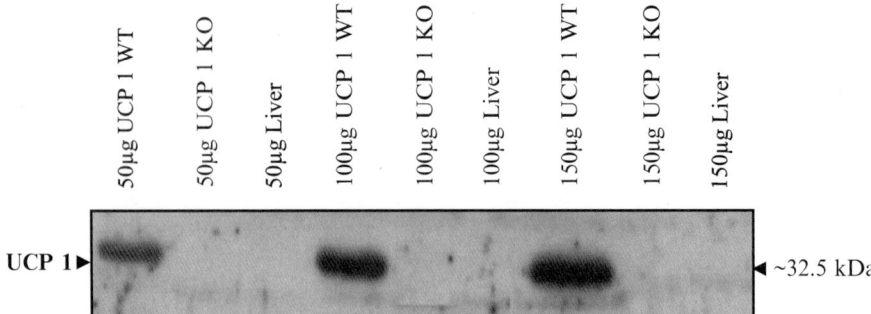

Fig. 3 UCP 1 peptide antibody (Eurogentec) detects UCP1 in BAT mitochondria from C57BL/6J wild type (WT) but not BAT from UCP 1 knockout (KO) mice. Varying concentrations of BAT (UCP1 present) and liver (UCP1 absent) mitochondria from C57BL/6J wild type and BAT mitochondria from UCP1 knockout mice were subjected to SDS-PAGE (12 % resolving gel) and immunoblotting with a 1:1,000 dilution of an anti-UCP 1 (Eurogentec) peptide antibody. A ~32.5 kDa denotes the presence of UCP 1

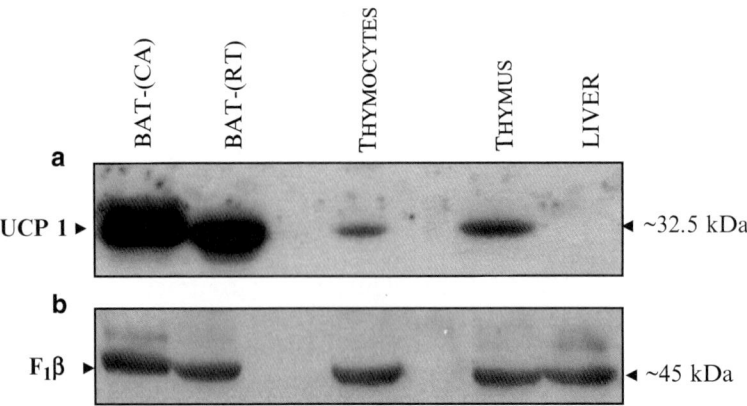

Fig. 4 UCP 1 peptide antibody (Eurogentec) detects UCP1 in BAT and thymus/thymocyte but not in liver of mitochondria from rat. Mitochondria isolated from (a) BAT of cold-acclimated (CA, 8 °C) rats (fed), (b) BAT of fed rats at room temperature (RT, 21 °C), (c) thymocytes of fasted rats at room temperature, (d) whole thymus of fasted rats at room temperature and (e) liver of fed rats at room temperature. 100 μg mitochondria were subjected to SDS-PAGE (12 % resolving gel) and immunoblotting with an anti-UCP 1 (Eurogentec) peptide antibody (**a**) and an antibody to the β-subunit of the F_1-ATPase (F_1 β) (**b**). All antibodies were used at a 1:1,000 dilution A ~33 and ~45 kDa band denotes the presence of UCP 1 and F_1β respectively

3.4 Measurement of Oxygen Consumption Rate

1. The Oroboros Respirometer was used to measure oxygen consumption by thymus mitochondria. The Respirometer contains 2×2 ml glass sealable chambers.

2. After addition of 0.1 mg mitochondria/ml incubation medium, final concentrations of 1 μg/ml oligomycin, 5 μM atractyloside and 10 mM glycerol-3-phosphate were added for oxygen consumption rate measurements by non-phosphorylating thymus mitochondria (*see* **Note 21**).

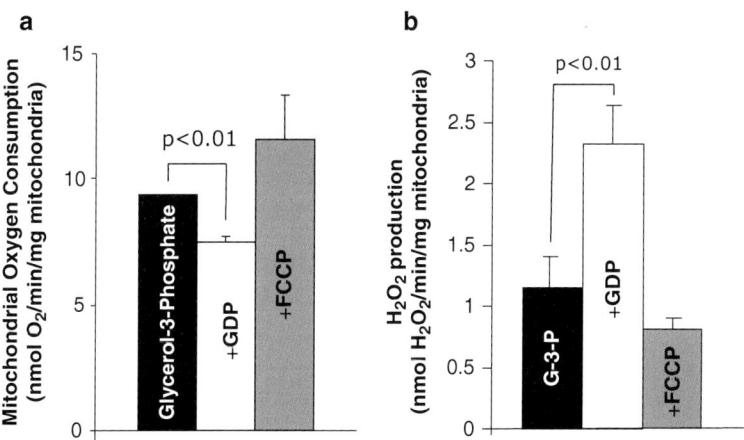

Fig. 5 Two Indirect assays of mitochondrial proton leak due to UCP1 in thymus mitochondria. Thymus mitochondria (100 μg/ml) incubated at 37 °C in 120 mM KCl, 5 mM Hepes-KOH pH 7.4, 1 mM EGTA, 1 μg/ml oligomycin, 5 μM atractyloside, 0.1 % de-fatted BSA and then substrate was added (10 mM glycerol-3-phosphate) (**a**) Steady-state oxygen consumption rates were obtained in the Oroboros Oxygraph Respirometer. 1 mM GDP was added to inhibit oxygen consumption due to UCP1 and 0.5 μM FCCP was added to achieve maximal uncoupled rate. (**b**) 5 μM Amplex Red, 10 U/ml horseradish peroxidase and 30 U/ml superoxide dismutase were added to the medium. Steady-state reactive oxygen species production was measured using a Perkin Elmer LS 55 fluorometer with excitation wavelength set at 570 ± 8 nm and emission wavelength set at 585 ± 4 nm as it detected an increase in fluorescence of Resorufin from Amplex Red. H_2O_2 production was measured for 250 s, whereupon 0.5 μM of the protonophore carbonyl cyanide p-(trifluoro-methoxy) phenylhydrazone (FCCP) was added and H_2O_2 production was measured for a further 250 s. Fluorescence was calibrated using known amounts of H_2O_2 each experimental day. Data in both graphs, is expressed as mean ± S.E.M. of at least three experiments, each experiment performed in triplicate. The mean values were compared using an unpaired (two-tailed) student's t-test. A probability p-value is thus obtained indicating whether the means are significant ($P < 0.05$)

3. Non-phosphorylating thymus mitochondrial oxygen consumption rates were measured when a stable signal was established (within a minute) for up to 2–3 min (Fig. 5a).

4. GDP (1 mM final concentration) was added to the chamber to inhibit oxygen consumption due to proton leak by UCP1.

5. The protonophore carbonyl cyanide p-(trifluoro-methoxy) phenyl-hydrazone (FCCP) (0.5 μM final concentration) was added to the chamber to determine maximal "uncoupled" oxygen consumption rate.

3.5 Measurement of Mitochondrial Reactive Oxygen Species (H₂O₂) Generation

1. A Perkin Elmer LS 55 fluorometer with excitation set at 570 ± 8 nm and emission at 585 ± 4 nm was used to detect fluorescence of Resorufin.

2. H_2O_2 generation, detected by the Amplex Red conversion to Resorufin, was essentially as described by Dlasková et al. [10] (*see* **Note 22**).

3. After addition of 0.1 mg mitochondria/ml incubation medium, final concentrations of 1 µg/ml oligomycin, 5 µM atractyloside, 10 mM glycerol-3-phosphate, 5 µM Amplex Red, 10 U/ml horseradish peroxidase and 30 U/ml superoxide dismutase were added to detect reactive oxygen species production by thymus mitochondria (Fig. 5b).

4. GDP (1 mM final concentration), an inhibitor proton leak due to UCP1 and which increases reactive oxygen species production, was added to mitochondrial suspensions in fluorimeter cuvettes.

5. The protonophore carbonyl cyanide p-(trifluoro-methoxy) phenyl-hydrazone (FCCP) (0.5 µM final concentration) which completely uncouples mitochondria and decreases reactive oxygen species production, was added to mitochondrial suspensions in fluorimeter cuvettes.

4   Notes

1. Mitochondria from whatever source are traditionally isolated in non-ionic media and for good reason. The functionality of the resulting mitochondrial fraction is preserved in a preparation from a non-ionic medium when compared preparations with an ionic medium. In addition, the inclusion of the calcium chelator ethylene glycol-bis (2-aminoethylether)-N,N,N',N'-tetraacetic acid (EGTA) results in a good quality mitochondrial preparation.

2. The tin foil is to allow you to place the pestles in the ice without ice contacting the pestle. No ice/water should get onto the pestles or in the homogenizer.

3. Young animals are selected as they have large thymi when compared with older animals. The thymus begins to atrophy once adolescence begins.

4. Best to prepare this fresh each time.

5. Due to the obnoxius odor of mercaptoethanol, ours and other laboratories are now using the less ordorous reductant dithiothreitol.

6. PVDF membranes are more durable and have higher binding capacities when compared to nitrocellulose membranes (150–160 µg/cm^2 vs. 80 µg/cm^2). Thus they have higher sensitivities and are the preferred choice for hydrophobic (i.e., membrane) proteins like UCP1. PVDF membranes requires a brief "wetting" step with methanol.

7. We used the same solution for Blocking and Probing.

8. Semidry blotting is easier, quicker and uses less buffer than wet blotting. Low molecular weight proteins transfer with greater

reproducibility using semidry blotting whereas high molecular weight proteins transfer better with wet transfer. UCP1 is in the medium range and thus either method can be used.

9. Ponceau S is a diazo dye used to determine the presence of protein on nitrocellulose or PVDF membranes.

10. In this instance we used an antibody to the (145–158) peptide of UCP1 raised for us by Eurogentec. The antibody was affinity purified before use. However, we have also shown that both the Sigma and Calbiochem antibodies to the (145–158) peptide of UCP1 are sensitive and selective for UCP1 over UCP2, UCP3 and other mitochondrial transporters.

11. In this instance we used an antibody to the β-subunit of the F_1-ATP synthase from Neurospora crassa, a gift from a colleague (Dr. Matt Harmey, Department of Botany, University College Dublin, Ireland). However, other antibodies such as the antibody to the 142–171 region of the human β-subunit of the F_1-ATP synthase from Acris will work just as well.

12. The Oroboros Respirometer is a very sensitive Clark-type oxygen electrode with 2×2 ml glass sealable chambers. It is the most sensitive oxygen electrode on the market and further details can be obtained at the Oroboros website: http://www.oroboros.at/

13. Amplex Red is converted to Resorufin in the presence of hydrogen peroxide by the enzyme horseradish peroxidase included in the assay medium. As electrons "escape" from the electron transport chain they generate superoxide. Inclusion of superoxide dismutase in the assay medium converts superoxide to hydrogen peroxide and oxygen. So in short the Amplex Red assay indirectly measures superoxide production by mitochondria, by directly measuring hydrogen peroxide production rate, which in turn is detected by the increase in fluorescence due to the increase in abundance of Resorufin from Amplex Red. Steady-state hydrogen peroxide production was measured using a Perkin Elmer LS 55 fluorometer set to detect Resorufin with excitation wavelength at 570 ± 8 nm and emission wavelength set at 585 ± 4 nm.

14. Asphyxiation by carbon dioxide is an approved method for euthanasia of rats/mice. The method limits any damage or bleeding into the thymus when compared with other forms of euthanasia such as cervical dislocation.

15. The thymus is under the rib cage on top of the heart and is a "milky" white color whereas brown adipose tissue is "tan/brown" in color due to the large abundance of mitochondria and if present, is easily distinguishable and easily separated from the thymus (Fig. 2)

16. In Chappell [9] mitochondria are isolated by differential centrifugation. This method has been used extensively for isolation of liver mitochondria.

17. The pellet wash in 2 % (w/v) defatted BSA is essential for recovering GDP-sensitive thymus mitochondria. A pellet wash of 0.2 % (w/v) defatted BSA results in thymus mitochondria that are NOT sensitive to GDP, however brown adipose tissue mitochondria are sensitive to GDP at that lower defatted BSA concentration.

18. A subsequent wash of the pellet in the absence of defatted BSA is so as not to contaminate the protein determination assay with protein other protein from the original sample.

19. We have used both the Bicinchoninic Acid Assay described by Smith et al. [11] and the Markwell assay [12] as a means to quantify of protein concentration in the final mitochondrial suspension.

20. Allow space for stacking gel and overlay resolving gel with isobutanol or water

21. Atractyloside inhibits the adenine nucleotide carrier and oligomycin inhibits the ATPsynthase, insuring non-phosphorylating thymus mitochondria. Glycerol-3-phosphate is a substrate for mitochondrial glycerol-3-phosphate dehydrogenase which is located on the outer surface of the mitochondrial inner membrane.

22. Interestingly, direct superoxide detection by ethidium bromide was not increased on ablation of UCP1 when measured in isolated mouse brown adipose tissue mitochondria [10]. Whereas increased hydrogen peroxide production was detectable in the presence of horseradish peroxidise and superoxide dismutase in isolated mouse brown adipose tissue mitochondria from UCP1 knock-out mice when compared to wild-type mice when Amplex Red was used [10].

References

1. Klaus S, Casteilla L, Bouillaud F, Ricquier D (1991) The uncoupling protein UCP: a membraneous mitochondrial ion carrier exclusively expressed in brown adipose tissue. Int J Biochem 23:791–801

2. Carroll AM, Haines LR, Pearson TW, Brennan C, Breen EP, Porter RK (2004) Immunodetection of UCP1 in rat thymocytes. Biochem Soc Trans 32:1066–1067

3. Carroll AM, Haines LR, Pearson TW, Fallon PG, Walsh CM, Brennan CM, Breen EP, Porter RK (2005) Identification of a functioning mitochondrial uncoupling protein 1 in thymus. J Biol Chem 280:15534–15543

4. Porter RK (2006) A new look at UCP 1. Biochim Biophys Acta 1757:446–448

5. Adams AE, Hanrahan O, Nolan DN, Voorheis HP, Fallon P, Porter RK (2008) Images of mitochondrial UCP 1 in mouse thymocytes using confocal microscopy. Biochim Biophys Acta 1777:115–117

6. Adams AE, Kelly OM, Porter RK (2010) Absence of mitochondrial uncoupling protein 1 affects apoptosis in thymocytes, thymocyte/ T-cell profile and peripheral T-cell number. Biochim Biophys Acta 1797:807–816

7. Clarke KJ, Adams AE, Manzke LH, Pearson TW, Borchers CH, Porter RK (2012) A role

for ubiquitinylation and the cytosolic proteasome in turnover of mitochondrial uncoupling protein 1 (UCP1). Biochim Biophys Acta 1817:1759–1767

8. Clarke KJ, Porter RK (2013) Uncoupling protein 1 dependent reactive oxygen species production by thymus mitochondria. Int J Biochem Cell Biol 45:81–89

9. Chappell JB, Hansford RG (1972) In: Birnie GD (ed) Subcellular components: preparation and fractionation. Buttersworths, London, pp 71–91

10. Dlasková A, Clarke KJ, Porter RK (2010) The role of UCP 1 in production of reactive oxygen species by mitochondria isolated from brown adipose tissue. Biochim Biophys Acta 1797:1470–1476

11. Smith PK, Krohn RI, Hermanson GT, Mallia AK, Gartner FH, Provenzano MD, Fujimoto EK, Goeke NM, Olson BJ, Klenk DC (1985) Measurement of protein using bicinchoninic acid. Anal Biochem 150:76–85

12. Markwell MA, Haas SM (1978) A modification of the Lowry procedure to simplify protein determination in membrane and lipoprotein samples. Anal Biochem 87:206–210

13. Laemmli UK (1970) Cleavage of structural proteins during the assembly of the head of bacteriophage T4. Nature 227:680–685

Chapter 12

Exploring Liver Mitochondrial Function by ^{13}C-Stable Isotope Breath Tests: Implications in Clinical Biochemistry

Ignazio Grattagliano, Leonilde Bonfrate, Michele Lorusso, Luigi Castorani, Ornella de Bari, and Piero Portincasa

Abstract

The liver plays a pivotal role in a myriad of metabolic processes, including detoxification, glycolipidic storage and export, and protein synthesis. Breath tests employing ^{13}C as stable isotope have been introduced to explore such energy-dependent pathways involving mitochondrial function in the liver. Specific substrates are ketoisocaproic acid, methionine, and octanoic acid. In humans, the application of ^{13}C-breath tests ranges from nonalcoholic and alcoholic liver diseases to liver cirrhosis, hepatocarcinoma, preoperative and postoperative assessment of liver function, and drug-induced liver damage. Studying liver mitochondrial function by ^{13}C-breath tests represents a complementary tool to monitor complex metabolic processes in health and disease.

Key words Breath test, Hepatic mitochondrial function, Hepatocellular carcinoma, Ketoisocaproic acid, Liver diseases, Liver steatosis, Methionine, Octanoic acid

1 Introduction

Mitochondria are intracellular organelles that act as source of energy for the cells in the body. Hepatocytes contain a large number of mitochondria (1,000–2,000 per cell) which participate in the intense metabolic processes occurring anytime in the liver during the degrading pathways involving carbohydrates, proteins, lipids, and xenobiotics. Mitochondrial dysfunction greatly contributes to the onset and progression of several chronic liver diseases [1], and mitochondrial abnormalities are reported in more advanced forms of liver diseases, including liver cirrhosis, as well as in drug-induced liver injury [2–4]. Chronic liver diseases, whether auto-immune, alcoholic, viral, or metabolic, are common conditions worldwide with impact on morbidity, mortality, quality of life of populations, and costs for the health care system [5]. Standard monitoring of patients with chronic liver disease is based on serum transaminases levels, imaging, and liver biopsy with histology. As

Carlos M. Palmeira and Anabela P. Rolo (eds.), *Mitochondrial Regulation*, Methods in Molecular Biology,
vol. 1241, DOI 10.1007/978-1-4939-1875-1_12, © Springer Science+Business Media New York 2015

noninvasive diagnostic tool, liver ultrasonography and, recently, ultrasound-based transient elastography and some algorithms based on serum parameters [6, 7] are employed to obtain a raw picture of liver disease status [8]. To date, no specific easily available blood test provides information on liver mitochondrial status. Mitochondrial metabolic processes are often investigated in isolated organelles or even in mitochondrial fractions or cell culture. Measurements, however, are complex in whole biological systems and make the comparison between different models and in vivo somewhat difficult. One example is the study of the effects of several xenobiotics, including drugs with known liabilities [9, 10]. The investigation becomes difficult when considering whole-body drug metabolism and conversion pathways into more or less reactive metabolites. Impaired mitochondrial function likely spreads metabolic markers into the systemic circulation, for example leading to abnormal acetoacetate/β-OH-butyrate ratio [11, 12]. Most of these parameters, however, are not easy to monitor and are not peculiar to specific functional abnormalities in the liver. Novel fingerprints of mitochondrial function are likely to emerge with the increasing use of metabolomics [13], a metabolite network aimed to quantify all metabolites in a cellular system under defined states. Metabolomics would allow the dynamic study of genetic perturbations [14]. Abnormal steady state concentrations of mitochondrial metabolites are not simple to interpret. Complex metabolic processes, in fact, might be influenced also by the influx of unknown precursors following raveled metabolic pathways. Methods for assessing and monitoring mitochondrial function and especially early mitochondrial dysfunction in the clinical setting might therefore provide additional diagnostic and prognostic information, and clues to novel therapeutic strategies. Problems related to the study of mitochondrial function could be potentially circumvented by the use of dynamic tests based on hepatic clearance. Dynamic tests such as breath tests (BTs), by measuring the grade of mitochondrial injury would contribute to exploring the progression of liver diseases in the clinical setting and in a noninvasive way [12]. In this scenario, known quantities of specific traced substrates can be quantitatively and selectively measured after either oral or intravenous administration. The easiest way is to measure gas in expired air, which represents the basis of dynamic BTs. For mitochondria, most used ^{13}C-substrates include ^{13}C-ketoisocaproic acid (^{13}C-KICA), ^{13}C-methionine, and ^{13}C-octanoate. Although studies appear still experimental, BTs could serve as a valuable supporting diagnostic tool in patients with specific liver diseases, to minimize the need for invasive procedures, i.e., liver biopsies (Table 1)

Table 1
Liver mitochondrial breath tests: substrates and evidence for potential clinical applications

Substrate	Clinical conditions	Information	Reference
KICA	ALD	Diagnosis of acute alcohol consumption (even low-moderate doses)	[26]
	ALD	Discrimination between chronic alcohol consumption and nonalcoholic chronic liver disease	[19]
	ALD	Monitoring and ascertaining of alcohol withdrawal	[19]
	NAFLD	Discrimination between simple steatosis and steatohepatitis (NASH) and low-grade and high-grade fibrosis	[22]
	HCC	Additional diagnostic parameter and prediction of tumor recurrence after local treatment	[23]
	Drugs	Evaluation of acute drug toxicity	[26]
Methionine	Liver cirrhosis	Discrimination between different degree of chronic liver damage	[29]
	ALD	Diagnosis of acute alcohol ingestion	[30]
	NAFLD	Discrimination between simple steatosis and NASH	[54]
	HCV	Discrimination between HCV infected patients and healthy subjects and toxicity of pegylate interferon plus ribavirin treatment	[42]
	Drugs	Evaluation of chronic drug toxicity	[31]
	Friedriech's ataxia	Diagnosis of neurological disorders	[34]
Octanoate	NAFLD	Evaluation of altered lipid metabolism	[40]

ALD alcoholic liver disease, *HCC* hepatocellular carcinoma, *HCV* hepatitic C virus, *KICA* α-ketoisocaproic acid, *NAFLD* nonalcoholic fatty liver disease, *NASH* nonalcoholic steatohepatitis

2 Materials

Physiologically, if a test substrate undergoes mitochondrial catabolism, CO_2 will be produced. If the carbon atom is labeled (*C), the appearance of the labeled $*CO_2$ in expired air will reflect the mitochondrial clearance of that given substrate. The general principle of the BT is based on such assumption [12]. The radioactive ^{14}C isotope has been used in the past, with some limitations. Instead, the stable, nonradioactive, environmental isotope ^{13}C is currently used. ^{13}C accounts for about 1.1 % of all natural carbon on Earth, and is found in plants and food chain. Several metabolic processes in hepatocytes may be explored by ^{13}C-labeled substrates and BTs (Fig. 1). The most used ^{13}C substrates for the study of mitochondrial function are ^{13}C-KICA, ^{13}C-methionine, and ^{13}C-octanoate (Fig. 2a–c). Benzoic acid undergoes glycine conjugation, and although interesting, has been used only in experimental animal models of liver cirrhosis [15].

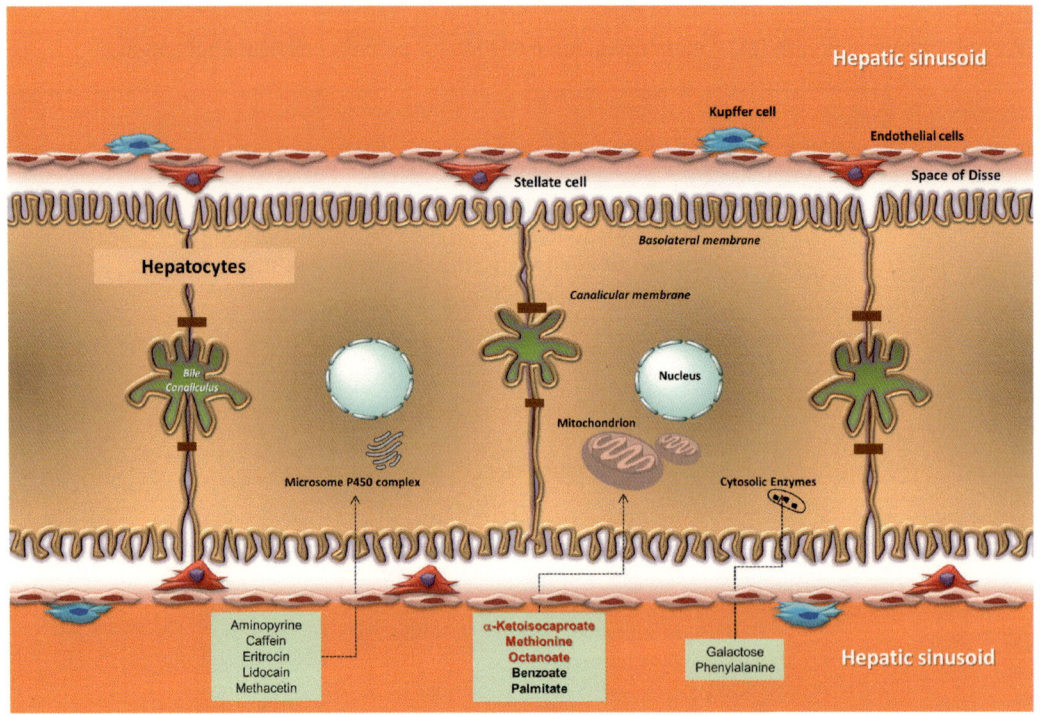

Fig. 1 Sites where metabolic processes may be explored by breath test in hepatocytes. In particular, [13C-α-ketoisocaproic] acid, [13C]-methionine, and [13C]-octanoate are the three substrates more widely employed for the assessment of mitochondrial function (see text for details)

2.1 ¹³C-KICA BT

KICA (MW: 130.141800 g/mol, MF: $C_6H_{10}O_3$, IUPAC name: 4-methyl-2-oxopentanoic acid) is an intermediate in the metabolism of leucine (Fig. 2a).

1. The decarboxylation of KICA is a specific function of mitochondria, as shown in experimental models [16], in isolated mitochondria [17], in healthy subjects with therapeutic doses of acetyl salicylic acid or with low ethanol intake, and in a number of liver diseases.

2. The generation of labeled CO_2 following administration of labeled KICA reflects mitochondrial function, particularly if the major competing pathway for the KICA elimination, the transamination to leucine, is suppressed by the concomitant administration of leucine (*see* **Note 1**). There is no gender difference when results are corrected for body composition [18].

3. KICA decarboxylation is decreased in alcoholics compared with patients with nonalcoholic fatty liver disease (NAFLD) and controls [19, 20], but not in subjects with chronic intake of alcohol or alcohol-induced steatosis [21]. KICA decarboxylation is defective in patients with advanced nonalcoholic

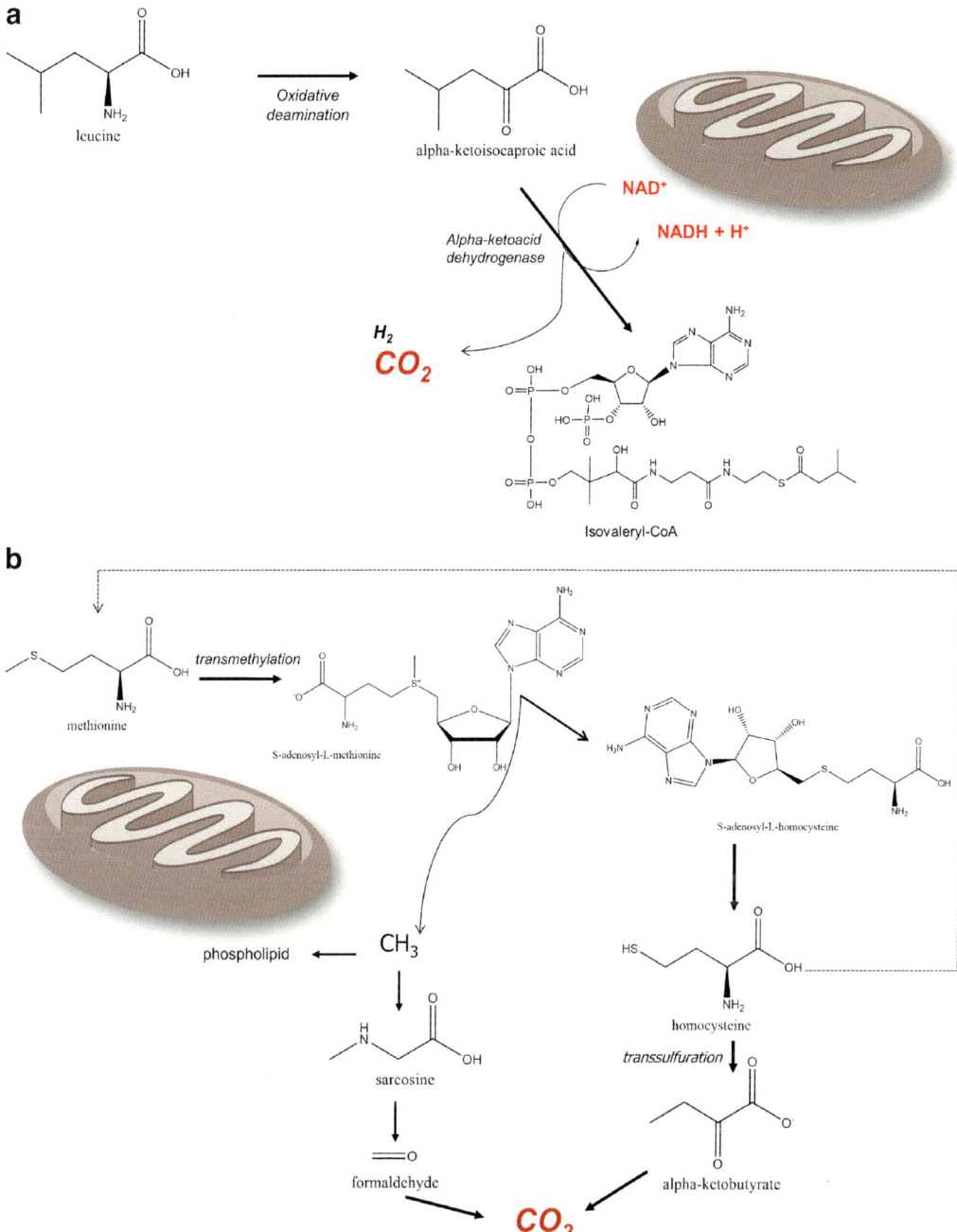

Fig. 2 Mitochondrial metabolism of (**a**) methionine; (**b**) α-ketoisocaproic acid; (**c**) octanoic acid. In all cases, CO_2 is produced. The principle of ^{13}C breath test relies on the use of ^{13}C-substrates which ultimately become donors of $^{13}CO_2$ following mitochondrial metabolism

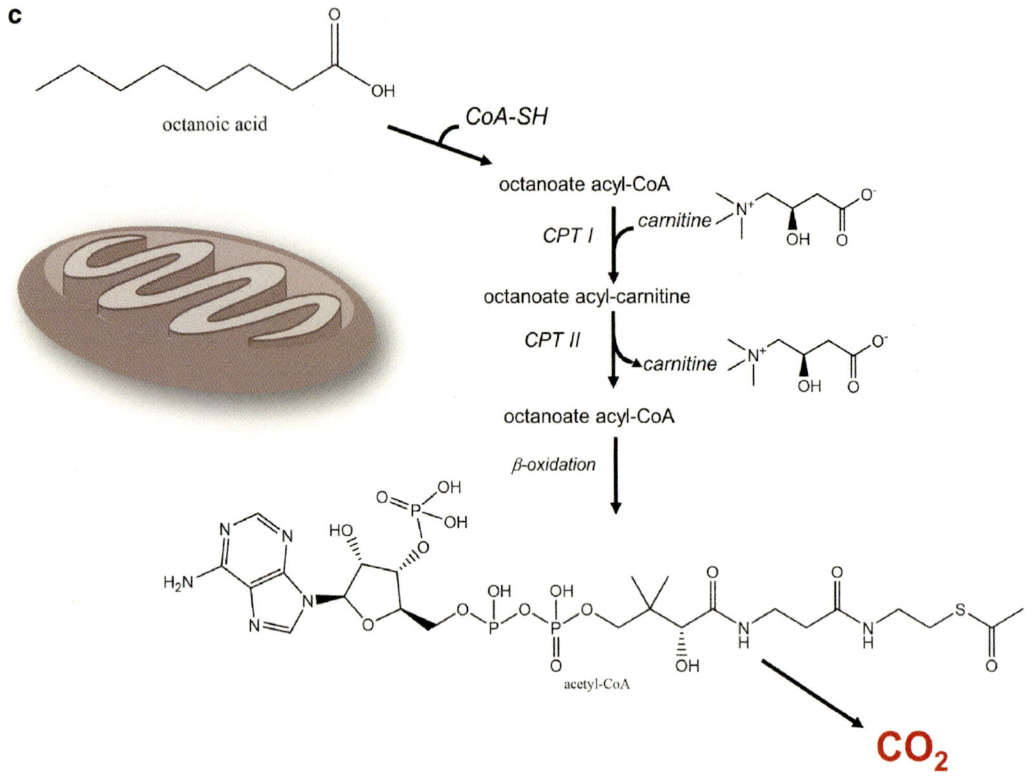

c

octanoic acid

CoA-SH

octanoate acyl-CoA

CPT I — carnitine

octanoate acyl-carnitine

CPT II — carnitine

octanoate acyl-CoA

β-oxidation

acetyl-CoA

CO_2

Fig. 2 (continued)

steatohepatitis (NASH) but not in those with simple steatosis [22], and inversely related to the extent of fibrosis resulting even better than measuring serum hyaluronate, especially in obese patients. KICA decarboxylation is also greatly decreased in cirrhotic patients with HCC compared with cirrhotic patients without HCC and identical Child–Pugh score [23] with further alterations just following treatment with radiofrequency ablation (RFA) and transarterial chemoembolization (TACE).

4. A recently novel application of ^{13}C-KICA BT has been described in a patient suffering from massive liver echinococcosis [24] (*see* **Note 2**). The test confirmed a slight mitochondrial malfunction (*see* **Note 3**).

5. Liver mitochondrial function is influenced by a number of drugs which enter mitochondria and accumulate, and interfering with respiratory complexes or electron transfer [25] (*see* **Note 4**). The test may be helpful to ascertain the integrity of these organelles before potentially toxic drugs are administered and to detect drug-induced mitochondrial damage before the appearance of symptoms in order to timely manage patients and prevent adverse effects. Examples are tacrolimus, aspirin [26], and ergot alkaloids [27].

2.2 ¹³C-Methionine BT

Methionine (MW: 149.21134 g/mol, MF: $C_5H_{11}NO_2S$, IUPAC name: (2S)-2-amino-4-methylsulfanylbutanoic acid) is an essential amino acid that plays a key role in various metabolic processes, including protein synthesis [28] (*see* **Note 5**).

1. If L-(1-¹³C)-methionine is administered as substrate, the metabolic process will be associated with production of labeled $^{13}CO_2$. Transmethylation of methionine results in the removal of the labeled methyl group if (methyl-¹³C)-methionine is used as substrate (Fig. 2b). The methyl group may in part be used for the synthesis of sarcosine which is oxidized to formaldehyde and production of CO_2 in mitochondria. Methionine differentially labeled in the methyl group and in position 1 can be used to study the complex metabolism of methionine [29].

2. For testing mitochondrial function, either L-(1-¹³C) methionine or (methyl-¹³C)-methionine may be used, but limitations apply (*see* **Note 6**).

3. Methionine BT has been used to assess mitochondrial function during acute intoxication and in chronic liver diseases: acute ethanol consumption impairs ¹³C-methionine decarboxylation in healthy volunteers with normal liver [30], whereas metabolism of methionine is decreased in patients with liver cirrhosis and especially in those with ethanol etiology [29], in those with biopsy-proven severe NAFLD in relation to the extent of steatosis [31], and in patients taking high dose valproic acid [31] or nucleoside analogs for the treatment of HIV [32].

4. Defective methionine BT has been reported also in hepatitis C infected cells [33], and in patients with Friedreich ataxia [34] an autosomal recessive degenerative disorder caused by loss of function mutations in the frataxin gene (*FXN* gene), located on chromosome 9q13 [35].

2.3 ¹³C-Octanoate BT

Octanoic acid (MW: 144.21144 g/mol, MF: $C_8H_{16}O_2$, IUPAC name: octanoic acid) is a medium chain fatty acid that enters mitochondria independently of the carnitine transport system.

1. Within mitochondria, octanoic acid undergoes β-oxidation which generates acetyl coenzyme A (AcCoA). AcCoA enters the Krebs cycle and is oxidized to CO_2 unless utilized for the synthesis of other energy-rich compounds (Fig. 2c).

2. As BT, ¹³C-octanoate is also used to assess the rate of gastric emptying to solids [36, 37] (*see* **Note 7**). ¹³C-Octanoate should also reflect hepatic mitochondrial function when gastric emptying and duodenal absorption are not severely delayed. Octanoic acid is therefore a potential substrate for noninvasive breath testing of hepatic mitochondrial beta-oxidation (*see* **Note 8**).

3. In animal models, the decarboxylation of octanoate was decreased in rats developing thioacetamide-induced acute hepatitis and liver cirrhosis [38].

4. In patients with NASH the oxidation of octanoate was either unchanged [39] or increased [40], and unchanged in those with early stage and advanced cirrhosis with and without porto-systemic shunt [41] (*see* **Note 9**).

3   Methods

With BTs, ^{13}C-labeled substrates are administered orally or intravenously. The most common way is by oral route (Fig. 3). The protocol steps are described in the following points [12, 23].

1. Overnight fasting subjects, 8–12 h (*see* **Note 10**).

2. The subject should rest in a quiet room and not smoking for at least 30 min before, and during the test (*see* **Note 11**).

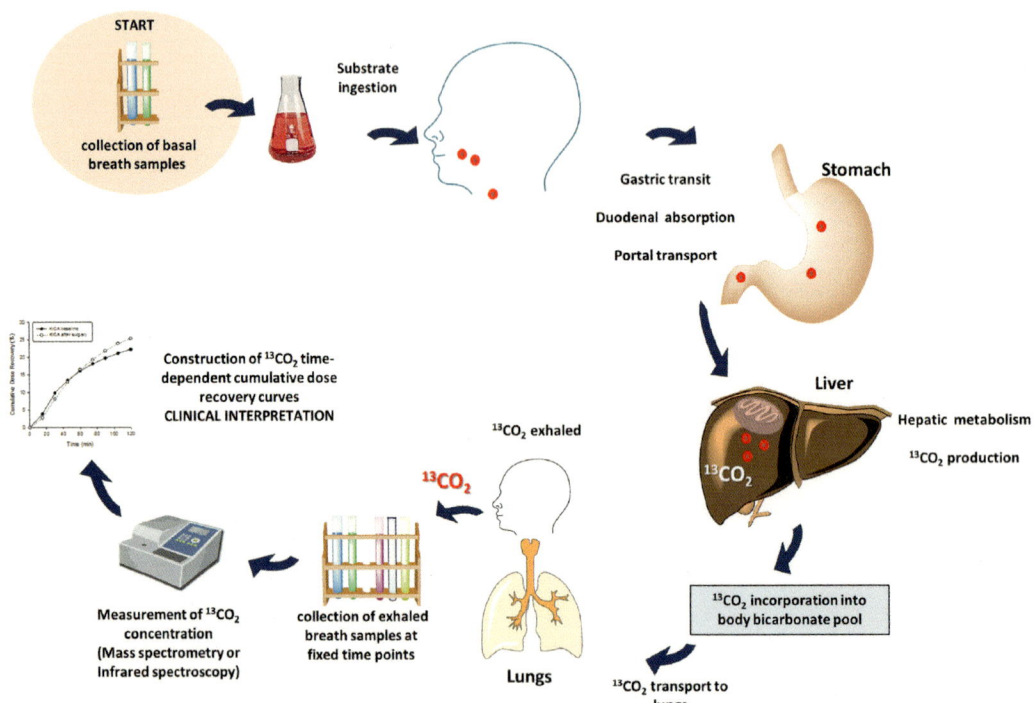

Fig. 3 Methodology of breath test analysis of ^{13}C-substrates for liver function. After ingestion of the given substrate, liver function is reflected by the concentration of exhaled $^{13}CO_2$, provided gastric emptying, duodenal absorption, portal transfer of the substrate to the liver, $^{13}CO_2$ distribution in the body compartments, and lung function are normal

3. The substrate is drunk by the patient within 1–3 min. For ^{13}C-KICA a dose of 1 mg/kg body weight plus 1 g unlabeled L-leucine is administered. ^{13}C-Methacetin is given at a dose of 1.5 mg/kg body weight. Substrates are dissolved in 100 mL of tap water (*see* **Note 12**).

4. Samples of expired air are collected into plastic bags (250–500 mL) or glass exetainer tubes, at baseline and thereafter at different time points, usually every 15 min up to 120 min (*see* **Note 13**).

5. The enrichment of expired ^{13}CO$_2$ is then analyzed by isotope ratio mass spectrometry or by infrared spectroscopy (i.e., IRMS; Helifan Plus, Fischer ANalysen Instrumente GmbH, Leipzig, Germany).

6. The rate of exhalation of ^{13}CO$_2$ at each time point is calculated from the measured increment in the isotopic abundance of ^{13}CO$_2$ (δ^{13}C$_{PDB}$), knowing the purity of the labeled compound and assuming a constant endogenous production of CO$_2$ of 300 mmol/m^2/h.

7. Results are expressed as percentage of the administered dose recovered per hour. The cumulative percentage of ^{13}CO$_{2\,in}$ breath is calculated as the area under curve (AUC) (*see* **Note 14**).

8. Results should always be compared with those obtained at each research center in a control group of healthy subjects (*see* **Note 15**).

9. Some limitations to the BT methodology apply (*see* **Note 16**).

10. Breath tests employing stable isotopes (^{13}C) might represent dynamic tools to investigate mitochondrial function in health and disease (*see* **Note 17**).

---

4 Notes

1. KICA decarboxylation is mainly influenced by the availability of NADH. For example earlier studies showed that ethanol given at 0.5 g/kg b.w. increases the availability of NADH and is associated with decreased KICA decarboxylation; by contrast, 1 g of aspirin decreases liver NADH concentrations and is associated with higher KICA decarboxylation [26].

2. Echinococcosis is an infection with the metacestode stage of the tapeworm Echinococcus, a member of the *Taeniidae* family. The most common forms *Echinococcus granulosus* and *Echinococcus multilocularis* are responsible for cystic echinococcosis and alveolar echinococcosis, respectively, and represent a reemerging infection with worldwide distribution [24].

3. The mitochondrial dysfunction was confirmed by a lower percentage of Cumulative Dose Recovery (CPDR) of 22 % (normal: CPDR within 120 min ≥ 23 % normal). By contrast, ^{13}C-methacetin breath test (as index of liver microsomal function) was normal. Likely, the slightly impaired mitochondrial function might be secondary to the "mass effect" of Echinococcus in the liver and/or to release of local hepatotoxic substances. In this patient, the test returned to normal weeks after hepatic surgery and excision of the cyst.

4. For example, nucleoside analogs once incorporated into mitochondrial DNA, inhibit c-DNA polymerase and hinder the replication process [32]. These drugs are widely used in transplanted patients or in HIV and HBV infected subjects. In any case, this is an intriguing question since it seems that also some viral infection itself (HIV, HCV) may impair mitochondrial function. In fact, mitochondrial function was found to be imbalanced in HCV infected cells [33] and in patients [42]. Several xenobiotics, including clinically relevant drugs have been described to cause excessive activation of mitochondrial permeability transition pore. They include acetaminophen [43], N-nitrosofenfluramine [44], or salicylate [45]. Nimesulide has shown these effects in vitro [46]. Aspirin, ibuprofen (nonsteroidal anti-inflammatory drugs), amiodarone (antiarrhythmic agent), and valproate (an anticonvulsant, histone deacetylase inhibitor) inhibit mitochondrial fatty acid β-oxidation [25, 47].

5. Exogenous methionine is used for protein synthesis or enters the methionine cycle in which it is transformed into S-adenosylmethionine, the main biological methyl donor. After donation of the methyl group, the resulting S-adenosylhomocysteine is hydrolyzed to homocysteine, which either undergoes transsulfuration or is remethylated to methionine. The transsulfuration step leads to the formation of α-ketobutyrate which enters mitochondria and ultimately undergoes decarboxylation.

6. The complex metabolic pathways of methionine render the interpretation of the BT results and especially the comparisons across studies quite difficult. This is especially true when differently labeled methionine is used, and more than one $^{13}CO_2$ is formed from either L-(1-^{13}C) or (methyl-^{13}C)-methionine [48]. Moreover, in some studies intravenous rather than oral methionine has been used and this makes comparisons even more difficult [49].

7. Currently, 100 mg octanoate labeled with ^{13}C is used as substrate already incorporated in a solid muffin test meal for the

study of gastric emptying. The test meal is ingested together with a glass of water. In a recent validation study [37], the test meal contained a total energy intake of 378 kcal with 57 g carbohydrate (61 %), 14 g fat (33 %), and 6 g protein (6 %) (SOFAR, Milan, Italy). The retention of ^{13}C-octanoic acid to the muffin was evaluated by in vitro incubation studies [50]. A similar meal consisting of three small muffins to which 50–100 mg ^{13}C-octanoic acid is provided within a soluble capsule taken orally, is being validated. (Portincasa et al., internal file and MD Thesis, De Luca et al. University of Bari, October 2013; manufacturer AB Analitica SrL, Padua, Italy). For gastric emptying studies breath samples are taken at baseline and every 30 min during 4 h. The software for calculation of results is provided by the manufacturer.

8. The substrate (98 % APE) can be obtained by local manufacturers involved in stable isotope distribution or from few international companies (e.g., Cambridge Isotope Laboratories, Andover, MA, USA). For clinical experimental application, a specific authorization of local ethical committees is necessary, in spite of noninvasiveness of the substrate. Informed consent needs to be provided, and includes full explanation of potential risks (none), costs, benefits and study protocol. The informed consent must be provided *before* the test starts, and is dated and signed by the subject undergoing the test, the physician in charge, and a witness.

9. Such apparently discrepant results with octanoate might be due to subtle differences in metabolic pathways, substrates employed or by extra-hepatic mitochondrial oxidation of octanoate. Unfortunately, a comparison of different substrates and BTs in the same group of subjects/patients has not been performed, so far.

10. The clinical history or previous investigations should suggest that gastric emptying is preserved and not severely delayed. The latter condition (i.e., gastroparesis, major motility defects, inflammation, malignancies, etc.) might interfere with the delivery of the substrate to the duodenum, and make interpretation of the results difficult.

11. Avoiding movement and smoking will minimize variations in endogenous CO_2 production due to physical activity or combustion.

12. The substrate is flavorless and the small volume in which it is dissolved allows a quick ingestion with prompt initiation of gastric emptying process. This is due to the liquid composition of the sample.

13. In normal conditions, a total of nine samples are taken for a 15-min sampling during 2 h (i.e., at time 0, 15, 30, 45, 60, 75, 90, 105, 120 min). Depending on local technology and facilities, the investigator may prefer the use of plastic bags when the equipment is arranged to measure few samples (i.e., four at the same time for the Helifan equipment). Special glass tubes (exetainers) with a plastic/rubber cap are useful when several samples are planned on the same day or the automatic sampling system is available. One advantage of BTs is that if the patient is appropriately instructed, samples can be collected at home, in the ward or in the outpatient clinics from several patients. Bags and exetainers are tightly closed and samples are measured within 24 h, where the equipment is available. This is the concept of "field test" by using BTs.

14. Specific softwares are available for automatic calculation and construction of $^{13}CO_2$ exhalation versus time curves. The rate of exhalation of $^{13}CO_2$ at each time point is calculated from the measured increment in the isotopic abundance of $^{13}CO_2$ ($\delta13CPDB$), the known purity of the labeled compound and an assumed constant endogenous production of CO_2 (300 mmol/m^2/h), and is expressed as the percentage of the administered dose exhaled/h. The cumulative recovery of $^{13}CO_2$ in breath is calculated as the AUC (area under the curve) of the $^{13}CO_2$ exhalation rate compared with the time curve determined by linear interpolation using the trapezoidal rule [51, 52].

15. It is advisable that for clinical studies, the referral center will recruit a suitable number of sex-, age-, body size-matched control subjects with no evidence of gastrointestinal motility defects or major diseases or drug ingestion, or liver abnormalities. Results from such healthy controls (i.e., expressed as the cutoff above the mean + 2 standard deviations), will be compared with those obtained from the group of patients or from a single patients, looking for example at the difference at different time-points (e.g., 30, 60, 90, 120 min) or as overall AUC.

16. The use of stable isotopes allows the test repetition to monitor the course of a disease or the effect of a given treatment, and notably can be performed also in infants and pregnant women. Limited availability and costs of the equipment, costs of substrates, and need for experienced operators for a rather time-consuming test represent some of the drawbacks of this methodology. Almost all substrates used are naturally occurring compounds, are administered at very low doses, and toxicity is unlike to occur. Some of these substrates undergoing liver metabolism at various intracellular levels (Fig. 3) have been

utilized in clinical investigations although none of them has been currently officially approved for clinical use [12, 22, 23]. Thus, for clinical studies, authorization from the local ethical committees is required [12].

17. Variable results have been reported with ^{13}C-BTs when mitochondrial function is investigated in vivo, likely due to factors influencing mitochondrial performances. Nevertheless, a number of clinically relevant conditions can be investigated by ^{13}C-BT tests. Marked steatosis and NASH, as well as ethanol consumption are associated with liver mitochondrial dysfunction with KICA and methionine BT. The predictive value of BTs, however, does not allow to make clinical decisions based on BT results. The interpretation of mitochondrial BT is not straightforward and confounding variables have to be carefully considered when assessing the pathophysiological relevance and quantitative implications of the results. Orally administered substrates encounter influencing factors such as gastric emptying, bioavailability, and hepatic first pass metabolism. Competing pathways of elimination and metabolism of the test compounds and "competing" mitochondria, i.e., mitochondrial metabolism in organs other than the liver, may influence the test results. When substrates that also occur endogenously are used, the pool of endogenous substrate will have an impact on the concentration of the tracer and thereby influence the amount of tracer that is metabolized independently of mitochondrial function. Finally, the concentration of labeled CO_2 in breath will depend on the amount of CO_2 ending up in breath rather than in circulating or renally excreted bicarbonate and the endogenous production of unlabeled CO_2, which can vary substantially among subjects [53]. These variables influencing the test result are difficult to control. Nevertheless, BTs still have clinical utility in regard to diagnosis, prognosis or control of treatment effects if they can be adequately validated. Finally, BTs with new substrates investigating specific mitochondrial functions are warranted.

Acknowledgements

This work is partly supported by grants from the University of Bari (ORBA09XZZT, ORBA08YHKX). We are indebted to Carlos Palmeira, Paulo Oliveira and Catia Diogo (Coimbra University, Portugal) for longstanding scientific discussions and for sharing collaboration. We thank Rosa De Venuto, Paola De Benedictis, and Michele Persichella for skillful technical support.

References

1. Lee WS, Sokol RJ (2007) Liver disease in mitochondrial disorders. Semin Liver Dis 27: 259–273

2. Russmann S, Kullak-Ublick GA, Grattagliano I (2009) Current concepts of mechanisms in drug-induced hepatotoxicity. Curr Med Chem 16:3041–3053

3. Grattagliano I, Bonfrate L, Diogo CV et al (2009) Biochemical mechanisms in drug-induced liver injury: certainties and doubts. World J Gastroenterol 15:4865–4876

4. Grattagliano I, Portincasa P, D'Ambrosio G et al (2010) Avoiding drug interactions: here's help. J Fam Pract 59:322–329

5. Sherlock S, Dooley J (2002) Diseases of the liver and biliary system. Blackwell Science, Oxford

6. Ziol M, Handra-Luca A, Kettaneh A et al (2005) Noninvasive assessment of liver fibrosis by measurement of stiffness in patients with chronic hepatitis C. Hepatology 41:48–54

7. Curry MP, Afdhal NH (2013) Tests used for the noninvasive assessment of hepatic fibrosis. UpToDate Version 20.0

8. Castera L, Vergniol J, Foucher J et al (2005) Prospective comparison of transient elastography, fibrotest, APRI, and liver biopsy for the assessment of fibrosis in chronic hepatitis C. Gastroenterology 128:343–350

9. Nadanaciva S, Will Y (2011) New insights in drug-induced mitochondrial toxicity. Curr Pharm Des 17:2100–2112

10. Pereira CV, Nadanaciva S, Oliveira PJ et al (2012) The contribution of oxidative stress to drug-induced organ toxicity and its detection in vitro and in vivo. Expert Opin Drug Metab Toxicol 8:219–237

11. Grattagliano I, de Bari O, Bernardo TC et al (2012) Role of mitochondria in nonalcoholic fatty liver disease-from origin to propagation. Clin Biochem 45:610–618

12. Grattagliano I, Lauterburg BH, Palasciano G et al (2010) (13)C-breath tests for clinical investigation of liver mitochondrial function. Eur J Clin Invest 40(9):843–850

13. Masuo Y, Imai T, Shibato J et al (2009) Omic analyses unravels global molecular changes in the brain and liver of a rat model for chronic Sake (Japanese alcoholic beverage) intake. Electrophoresis 30:1259–1275

14. Griffin JL, Nicholls AW (2006) Metabolomics as a functional genomic tool for understanding lipid dysfunction in diabetes, obesity and related disorders. Pharmacogenomics 7: 1095–1107

15. Krahenbuhl L, Ledermann M, Lang C et al (2000) Relationship between hepatic mitochondrial functions in vivo and in vitro in rats with carbon tetrachloride-induced liver cirrhosis. J Hepatol 33:216–223

16. Michaletz PA, Cap L, Alpert E et al (1989) Assessment of mitochondrial function in vivo with a breath test utilizing alpha-ketoisocaproic acid. Hepatology 10:829–832

17. Grattagliano I, Vendemiale G, Lauterburg BH (1999) Reperfusion injury of the liver: role of mitochondria and protection by glutathione ester. J Surg Res 86:2–8

18. Berthold HK, Giesen TA, Gouni-Berthold I (2009) The stable isotope ketoisocaproic acid breath test as a measure of hepatic decarboxylation capacity: a quantitative analysis in normal subjects after oral and intravenous administration. Liver Int 29:1356–1364

19. Lauterburg BH, Liang D, Schwarzenbach FA et al (1993) Mitochondrial dysfunction in alcoholic patients as assessed by breath analysis. Hepatology 17:418–422

20. Witschi A, Mossi S, Meyer B et al (1994) Mitochondrial function reflected by the decarboxylation of [13C]ketoisocaproate is impaired in alcoholics. Alcohol Clin Exp Res 18: 951–955

21. Bendtsen P, Hannestad U, Pahlsson P (1998) Evaluation of the carbon 13-labeled Ketoisocaproate breath test to assess mitochondrial dysfunction in patients with high alcohol consumption. Alcohol Clin Exp Res 22:1792–1795

22. Portincasa P, Grattagliano I, Lauterburg BH et al (2006) Liver breath tests non-invasively predict higher stages of non-alcoholic steatohepatitis. Clin Sci (Lond) 111:135–143

23. Palmieri VO, Grattagliano I, Minerva F et al (2009) Liver function as assessed by breath tests in patients with hepatocellular carcinoma. J Surg Res 157:199–207

24. Bonfrate L, Giuliante F, Palasciano G et al (2013) Unexpected discovery of massive liver echinococcosis. A clinical, morphological, and functional diagnosis. Ann Hepatol 12: 634–641

25. Pessayre D, Mansouri A, Haouzi D et al (1999) Hepatotoxicity due to mitochondrial dysfunction. Cell Biol Toxicol 15:367–373

26. Lauterburg BH, Grattagliano I, Gmur R et al (1995) Noninvasive assessment of the effect of xenobiotics on mitochondrial function in human beings: studies with acetylsalicylic acid and ethanol with the use of the carbon

13-labeled ketoisocaproate breath test. J Lab Clin Med 125:378–383

27. Danicke S, Diers S (2013) Effects of ergot alkaloids on liver function of piglets as evaluated by the (13)C-methacetin and (13)C-alpha-ketoisocaproic acid breath test. Toxins (Basel) 5:139–161

28. Storch KJ, Wagner DA, Burke JF et al (1988) Quantitative study in vivo of methionine cycle in humans using [methyl-2H3]- and [1-13C] methionine. Am J Physiol 255:E322–E331

29. Russmann S, Junker E, Lauterburg BH (2002) Remethylation and transsulfuration of methionine in cirrhosis: studies with L-[H3-methyl-1-C]methionine. Hepatology 36:1190–1196

30. Armuzzi A, Marcoccia S, Zocco MA et al (2000) Non-Invasive assessment of human hepatic mitochondrial function through the 13C-methionine breath test. Scand J Gastroenterol 35:650–653

31. Spahr L, Negro F, Leandro G et al (2003) Impaired hepatic mitochondrial oxidation using the 13C-methionine breath test in patients with macrovesicular steatosis and patients with cirrhosis. Med Sci Monit 9:CR6–CR11

32. Milazzo L, Piazza M, Sangaletti O et al (2005) [13C]Methionine breath test: a novel method to detect antiretroviral drug-related mitochondrial toxicity. J Antimicrob Chemother 55:84–89

33. Li Y, Boehning DF, Qian T et al (2007) Hepatitis C virus core protein increases mitochondrial ROS production by stimulation of Ca2+ uniporter activity. FASEB J 21:2474–2485

34. Stuwe SH, Goetze O, Arning L et al (2011) Hepatic mitochondrial dysfunction in Friedreich ataxia. BMC Neurol 11:145

35. Durr A, Cossee M, Agid Y et al (1996) Clinical and genetic abnormalities in patients with Friedreich's ataxia. N Engl J Med 335: 1169–1175

36. Ghoos YF, Maes BD, Geypens BJ et al (1993) Measurement of gastric emptying rate of solids by means of a carbon-labeled octanoic acid breath test. Gastroenterology 104:1640–1647

37. Perri F, Bellini M, Portincasa P et al (2010) (13)C-octanoic acid breath test (OBT) with a new test meal (EXPIROGer): toward standardization for testing gastric emptying of solids. Dig Liver Dis 42:549–553

38. Shalev T, Aeed H, Sorin V et al (2010) Evaluation of the 13C-octanoate breath test as a surrogate marker of liver damage in animal models. Dig Dis Sci 55:1589–1598

39. Schneider AR, Kraut C, Lindenthal B et al (2005) Total body metabolism of 13C-octanoic acid is preserved in patients with non-alcoholic steatohepatitis, but differs between women and men. Eur J Gastroenterol Hepatol 17: 1181–1184

40. Miele L, Grieco A, Armuzzi A et al (2003) Hepatic mitochondrial beta-oxidation in patients with nonalcoholic steatohepatitis assessed by 13C-octanoate breath test. Am J Gastroenterol 98:2335–2336

41. van de Casteele M, Luypaerts A, Geypens B et al (2003) Oxidative breakdown of octanoic acid is maintained in patients with cirrhosis despite advanced disease. Neurogastroenterol Motil 15:113–120

42. Banasch M, Emminghaus R, Ellrichmann M et al (2008) Longitudinal effects of hepatitis C virus treatment on hepatic mitochondrial dysfunction assessed by C-methionine breath test. Aliment Pharmacol Ther 28:443–449

43. Jaeschke H, McGill MR, Ramachandran A (2012) Oxidant stress, mitochondria, and cell death mechanisms in drug-induced liver injury: lessons learned from acetaminophen hepatotoxicity. Drug Metab Rev 44:88–106

44. Nakagawa Y, Suzuki T, Kamimura H et al (2006) Role of mitochondrial membrane permeability transition in N-nitrosofenfluramine-induced cell injury in rat hepatocytes. Eur J Pharmacol 529:33–39

45. Trost LC, Lemasters JJ (1997) Role of the mitochondrial permeability transition in salicylate toxicity to cultured rat hepatocytes: implications for the pathogenesis of Reye's syndrome. Toxicol Appl Pharmacol 147: 431–441

46. Mingatto FE, dos Santos AC, Rodrigues T et al (2000) Effects of nimesulide and its reduced metabolite on mitochondria. Br J Pharmacol 131:1154–1160

47. Kass GE, Price SC (2008) Role of mitochondria in drug-induced cholestatic injury. Clin Liver Dis 12:27–51, vii

48. Candelli M, Miele L, Armuzzi A et al (2008) 13C-methionine breath tests for mitochondrial liver function assessment. Eur Rev Med Pharmacol Sci 12:245–249

49. Duro D, Duggan C, Valim C et al (2009) Novel intravenous (13)C-methionine breath test as a measure of liver function in children with short bowel syndrome. J Pediatr Surg 44:236–240, discussion 240

50. Bromer MQ, Kantor SB, Wagner DA et al (2002) Simultaneous measurement of gastric emptying with a simple muffin meal using [13C]octanoate breath test and scintigraphy in normal subjects and patients with dyspeptic symptoms. Dig Dis Sci 47:1657–1663

51. Dawson B, Trapp RG (2001) Basic & clinical biostatistics, vol 3. McGraw-Hill, New York

52. Hintze J (2013) NCSS 9. NCSS, LLC, Kaysville, UT www.ncss.com. Number Cruncher Statistical System (NCSS), Kaysville, UT

53. Winchell HS, Wiley K (1970) Considerations in analysis of breath 14CO2 data. J Nucl Med 11:708–710

54. Banasch M, Ellrichmann M, Tannapfel A et al (2011) The non-invasive (13)C-methionine breath test detects hepatic mitochondrial dysfunction as a marker of disease activity in non-alcoholic steatohepatitis. Eur J Med Res 16:258–264

Chapter 13

Following Mitochondria Dynamism: Confocal Analysis of the Organelle Morphology

Francesca R. Mariotti, Mauro Corrado, and Silvia Campello

Abstract

Mitochondria are highly dynamic organelles, whose morphology can vary from an elongated and interconnected network to fragmented units. In recent years, outstanding discoveries have linked mitochondrial morphology to the regulation of an increasing number of biological processes, such as bio-synthetic pathways, oxidative phosphorylation and ATP production, calcium buffering, and cell death. Here we describe two of the main methods used to analyze the mitochondrial length in fixed cells and the mitochondrial fusion rate in live cells. Moreover, we focus one of the protocols on T cells, as an example of non-adherent cells, which present some particularities and difficulties in the analysis of mitochondrial shape. We also discuss the main mouse models carrying a mitochondrial targeted fluorescent protein, an invaluable tool to deeply investigate in vivo mitochondrial morphology.

Key words Mitochondria, Morphology, Fusion and fission, Microscopy, Fluorescence

1 Introduction

1.1 Mitochondria Dynamics

Mitochondria play important roles in different cellular activities, spanning from energy production to cell death regulation. They constantly undergo structural transitions, passing from a fused and interconnected network to fragmented units [1]. The balance between fusion and fission events determines the mitochondrial shape. Moreover, mitochondria are actively transported throughout the cells and they can change their structure in response to physiological stimuli. For all these different aspects they are considered highly dynamic organelles.

The continuous shape changes have important effects on the morphology, function and distribution of mitochondria [2] and, at the same time, they affect various cellular processes [3, 4].

It is therefore clear that the balance between the different mitochondrial conformations must be tightly regulated and that any alterations can have detrimental effects on their stability and on cell survival. Indeed, mutations in genes regulating mitochondrial

Carlos M. Palmeira and Anabela P. Rolo (eds.), *Mitochondrial Regulation*, Methods in Molecular Biology, vol. 1241, DOI 10.1007/978-1-4939-1875-1_13, © Springer Science+Business Media New York 2015

Table 1
Schematic representation of the different approaches that will be described in the following paragraph for mitochondrial imaging

Approaches	Tools	Description
In vitro studies	*Antibodies* directed against mitochondrial proteins	Huge number of available antibodies; Limitation: in vitro studies with fixed cells (not dynamic) and low fluorescence/resolution
	Dyes (Rhodamine, Rosamine, JC-1, MitoTracker)	Analysis of mitochondria structure and status; Limitations: cytotoxicity, risk of artifacts impinging on the results
	Fluorescent proteins targeted to mitochondria (such as mitoGFP, mitoRFP, *photoactivatable fluorescent proteins*	Highly specific biosensors for mitochondria analyses, in particular for live imaging; bright fluorescence (which means images with good resolution); dynamic experiments; limitation: transfection/infection within cells not always feasible
In vivo studies	*Transgenic mice*	In vivo studies and ex vivo live cell imaging; tissue-dependent mitochondria visualization; experiments also with untransfectable cells. Limitations: not always possible to maintain and work with animals and to obtain transgenic mice

dynamics have been linked to the insurgence of cancer and neurodegenerative pathologies, such as Alzheimer [5], Parkinson [6], Autosomal Dominant Optic Atrophy (ADOA) [7], and Huntington's disease [8].

1.2 Studying Mitochondrial Dynamics: Different Approaches

The main need to study mitochondrial dynamics is to visualize the organelles with a resolution sufficiently high to appreciate differences in shape and/or size, also in small cells. Techniques for imaging mitochondria are continuously being improved to further unveil the mechanisms that regulate this process and to link the shape of these organelles to the different cellular processes (Table 1). Therefore, in this section we provide a brief description of the different approaches that are commonly adopted to visualize mitochondria and mitochondrial dynamics. First, we describe the tools available for mitochondrial imaging both in cultured cells and in transgenic mice. In the second paragraph, we instead provide detailed protocols to observe and measure mitochondrial morphology in different cell types (Table 1).

1.2.1 Mitochondria Imaging In Vitro

In Vitro/Ex Vivo Approaches

The simplest and easiest way to visualize mitochondria inside the cells is by Immunofluorescence, using the various commercially available antibodies against mitochondrial proteins. The incubation with secondary antibodies conjugated with fluorescent dyes allows mitochondrial visualization at the microscope. In this case, cells have to be fixed to a glass support and permeabilized before visualizing mitochondria, thus with the limitations that an in vitro experiment, with fixed cells, might have. Moreover, other technical limitations must also be taken in consideration, such as the stability and brightness of the fluorescence, the latter often so weak that it is easily bleached by the microscope lasers, and lost.

As an alternative, mitochondrial imaging can be performed taking advantage of commercially available dyes or constructs consisting of mitochondrial targeting sequences fused to fluorescent proteins [9]. Rhodamine, rosamine, JC-1, MitoTracker, and tetramethylrhodamine are some of the commercially available dyes. They are usually accumulated in the mitochondrial matrix, according to the plasma and the mitochondrial membrane potential, following Nernst Equation. Their fluorescence response to changes in mitochondrial membrane potential is then complicated by their peculiar tendency to undergo self-quenching. Moreover, at least in some conditions, changes in the mitochondrial potential and morphology can be mutually linked. Thus, for all these aspects, the use of these dyes might be improper or can give rise to artifactual results. Finally, cytotoxicity is another important limitation of these dyes that must be accurately considered before starting the experiment.

The best approach to properly visualize mitochondrial dynamics is the transfection within the cells of fluorescent proteins specifically targeted to mitochondria. Indeed, Fluorescent proteins (FP) and the most recently engineered photoactivatable FPs (PA-FP) are emerging as powerful tools for live imaging analyses [10]. The wild-type green fluorescent protein (GFP) has been modified to produce variant proteins excitable and emitting at different wavelengths, such as yellow (YFP), blue (BFP), and cyan (CFP). Genetic modifications of a red fluorescent protein extracted from *Discosoma striata* coral (DsRed), commonly referred to DsRed, allowed the formation of a monomeric RFP1 protein [11], thus avoiding the problems of aggregation frequently observed with "old generation" plasmids. A new version of DsRed has been further engineered, DsRed-Express2, which is stably expressed [12] without any sign of toxicity and phototoxicity, conversely detected with other proteins and that can be excited by both blue and green lasers. To further improve the fluorescence signals, enhanced variants of fluorescent proteins are widely used for imaging analyses [13]. However, one of the most important improvements in FP technology is the introduction of the photoactivatable-GFP (PA-GFP) [14].

To specifically visualize mitochondria, all the above mentioned plasmids have been modified fusing the fluorescent protein to a short sequence (targeting signal) specifically directed to the mitochondrial compartment, usually part of a protein resident in the mitochondria (i.e., one subunit of the human cytochrome c oxidase fused to the RFP protein, in the case of the pDsRed2-Mito vector).

Technically, fluorescent proteins give bright and stable fluorescent signals and allow live imaging analyses of mitochondrial dynamics without any risk of artifactual results. The only issue that might be considered as a technical limitation is the requirement of construct transfection/infection within the cells, sometimes unfeasible or even impossible in a number of cell types.

In Vivo Approaches

Together with the use of fluorescent proteins targeted to mitochondria, transgenic mice with mitochondrially targeted fluorescent proteins have been recently developed to allow and/or further improve the in vivo analyses of the mitochondrial network [15], also in specific tissues/organs or cell populations.

To this aim, different transgenic mice have been recently engineered. Yonekawa and colleagues have constructed a first type of transgenic mice, in which mito-EGFP cDNA is under the CMV early enhancer/β-actin (CAG) promoter to have a strong and ubiquitous expression in mitochondria of any tissues of the organism [15]. Transgenic mice have also been constructed to visualize mitochondria in a tissue-specific manner [16]. The Lox-Stop-Lox-Mito-YFP reporter mouse, engineered by Larsson and colleagues, contains a lox-flanked stop cassette upstream of the mito-YFP transgene; *cre*-dependent excision of the stop cassette allows YFP expression. According to the chosen *cre* promoters, YFP expression can be then restricted to specific tissues or cells [16].

Other examples of tissue-specific mouse lines recently obtained have enabled to perform mitochondrial live imaging in neurons [17] and microvascular endothelial cells [18].

Another important and innovative transgenic mouse has also been recently constructed [19]. Beside the flox-cre system to induce fluorescence in specific tissues/cell types, it contains the photo-convertible fluorescent protein Dendra2 fused to a mitochondrial matrix-target sequence. The power of this construct consists in the photo-switching ability of Dendra2. Indeed, Dendra2 can be very efficiently and irreversibly photoconverted from the green to the red fluorescent state, by light irradiation [20]; This enables visualization and tracking of both the activated red and green forms of the protein. In this way, mitochondria fusion/fission dynamic events can be directly and precisely measured in a subcellular population within tissues characterized by high cell diversity [19].

It can be concluded that the creation of transgenic mice expressing fluorescent proteins fused to mitochondrial sequences represent an important improvement, probably the most powerful tool for studying mitochondrial morphology, in tissue-specific and physiological dynamic conditions.

A list of most of the available transgenic mice for mitochondrial analyses can be found at the following website: http://www.jax.org/.

2   Materials

As explained in the previous paragraph, to analyze mitochondrial morphology it is often necessary to target a fluorescent protein to the organelle. In these protocols, we refer to mitochondrial targeted red fluorescent protein (mito-DsRed) and mitochondrial photoactivatable green fluorescent protein (mito-PA-GFP). For the imaging, a confocal microscope equipped with 413, 488, and 563 nm laser lines and with an incubation system (temperature, humidity, and gas control) will be necessary. It will be necessary to adjust the general protocols here presented for MEF, HEK293, and Jurkat cells to the specific experimental conditions.

2.1   Transfection of MEF or HEK293 Cells

1. TransFectin/Lipofectamine.
2. OptiMEM.
3. 13 mm glass coverslips.
4. PBS 1×.
5. DMEM with and w/o antibiotics.
6. Mito-DsRed plasmid.
7. Mito-PA-GFP.
8. Confocal microscope equipped with 413, 488, and 563 nm laser lines with temperature and humidity control.
9. Formaldehyde.
10. Anti-fade reagent.

2.2   Transfection of Jurkat Cells

1. Neon transfection system—100 μl kit (Invitrogen), or any other electroporator.
2. RPMI 1640 w/o antibiotics.
3. PBS w/o Ca^{2+} and Mg^{2+}.
4. Mito-DsRed plasmid.
5. Confocal microscope equipped with 563 nm laser line.
6. Special coverslips for not adherent cells.
7. Formaldehyde.
8. Anti-fade reagent.

3   Methods

3.1   Measuring Mitochondrial Length in Adherent Fixed MEF or HEK293 Cells

1. Place a round glass coverslip into a well plate and coat with poly-L-Lysine 50 μg/ml for 20 min.
2. Seed an appropriate number of cells in order to have 80 % confluence the day of transfection.

3. The day of transfection, wash the cells twice with PBS w/o Ca^{2+} and Mg^{2+}.

4. Add DMEM without antibiotics.

5. Perform transfection with mito-DsRed according to manufacturer's instructions (*see* **Note 1**).

6. 24 h after transfection perform the experiment of interest.

7. At the selected time points, fix the cells by washing with PBS and then incubating with formaldehyde 3.7 % (diluted in the same medium/solution) for 10–15 min at room temperature (*see* **Note 2**).

8. Place the coverslips on a glass slice and add anti-fade reagent.

9. At the confocal microscope, excite with the 563 nm laser.

10. Acquire single confocal fluorescent planes (trying to select a plane and fields where the mitochondrial network is well represented).

11. Perform the measurement of the mitochondrial length by using the software ImageJ (NIH) or similar software. In particular, measure the length of the mitochondrial major axis in at least ten mitochondria per cell and in a minimum of 50 cells per experiment.

12. Plot mitochondrial length in μm.

3.2 Studying Mitochondrial Morphology in Non-adherent Jurkat Cells (See Note 3)

1. Count the cells. Usually, up to 2–3 million of cells per transfection will be necessary (*see* **Note 4**).

2. Harvest the selected number of cells and wash once with PBS w/o Ca^{2+} and Mg^{2+} (*see* **Note 5**).

3. Perform transfection using an electroporator. In this protocol, we use the Neon Transfection System (Invitrogen) (*see* **Note 6**).

4. Prepare the Neon transfection system for the electroporation by placing tube and the electrolytic buffer according to manufacturer's instructions.

5. Set pulse, width, and number of pulse: 1410 mV, 30 ms, and 1 respectively (as suggested in manufacturer's instructions) for Jurkat cells (*see* **Note 7**).

6. Resuspend cells in resuspension buffer (R buffer)—100 μl per transfection—and add mito-DsRed plasmid (5 μg of DNA per transfection) (*see* **Note 8**).

7. Load cells in the tip, making sure that there are not air bubbles in the tip, and give the electroporation pulse.

8. After electroporation, seed the cells in a 24-well plate dish (one transfection per well) containing RPMI 1640 without antibiotics.

9. 24 h after transfection perform the experiment of interest.

10. At the selected time points, seed 4×10^4 cells on special coverslip for not adherent cells and let them adhere to the surface for 15 min at room temperature in the dark.

11. Fix cells with 3.7 % formaldehyde (diluted in the same medium/solution) 10 min RT.

12. After washing, add anti-fade reagent and close the coverslip.

13. At the microscope, excite with 563 nm laser lines and acquire series of z-stacks covering the entire volume of the cell (suggested z step 0.35 μm).

14. The z-stacks will be then projected as Z project of the standard deviation of the fluorescence intensity by using the ImageJ software to observe and analyze the mitochondrial network (*see* **Note 9**).

3.3 Measuring Mitochondrial Fusion Rate in Live Cells (HEK293 or MEF Cells)

1. Place a round glass coverslip into a well plate and coat with poly-L-Lysine 50 μg/ml for 20 min.

2. Seed an appropriate number of cells in order to have 80 % confluence the day of transfection.

3. The day of transfection, wash the cells twice with PBS w/o Ca^{2+} and Mg^{2+}.

4. Add DMEM without antibiotics.

5. Perform transfection with TransFectin/Lipofectamine with a combination of mito-DsRed and mito-PA-GFP plasmids (rate 1:2) according to manufacturer's instructions.

6. 24 h after transfection, place the glass coverslip in an appropriate coverslip holder.

7. Cover the cells with HBSS or (better) DMEM without phenol red and start the acquisitions at the microscope (*see* **Note 10**).

8. Activate the mito-PA-GFP fluorescence. Select a region of interest (ROI) in one z-plane and activate by using 100 % power of the 413 nm laser line.

9. After that, emission frames coming from the 488 and 563 nm excitation lines are acquired for at least 30 min.

10. Measure the standard deviation of the green fluorescence intensity in the entire cell normalized for the intensity of the mito-DsRed fluorescence by using the Multi measure plug-in of ImageJ (NIH) (*see* **Note 11**) [21, 22].

4 Notes

1. Perform transfection preferentially with TransFectin for MEF cells and Lipofectamine for HEK293 cells, in order to limit the toxic effect of the transfection reagent and thus limiting cell death, both affecting mitochondrial morphology.

2. Selecting the right fixative agent is important to limit perturbation of the mitochondrial network during the time of

fixation. We suggest using formaldehyde instead of paraformaldehyde or methanol in the study of mitochondria morphology.

3. Jurkat cells are described here as an example of not adherent cells. They are small round shape cells and their few mitochondria are dispersed all around the cell. Therefore, the observation of a single z plane in confocal analysis will not give any significant information about the mitochondrial morphology. The acquisition of a series of z-stacks covering the entire volume of the cell, and its projection or 3D reconstruction will be indeed necessary.

4. Be sure that cells are in the growing log phase by splitting the cells the day before the transfection to increase its efficiency (represented by this two parameters: cell viability and percentage of transfection).

5. Transfection will be performed by electroporation. Thus, it is imperative to reduce the amount of salts in solution by using PBS w/o Ca^{2+} and Mg^{2+}.

6. Electroporation as transfection method is compulsory in Jurkat cells. This method allows, indeed, a good transfection efficiency (extremely limited with other methods) counteracted however by high rates of mortality. In this protocol we use the microporator Neon Transfection System (Invitrogen), that strongly limits cell death compared to other electroporators, and we give brief indications on how to use it. Any electroporator can be used, by following manufacturer's instructions.

7. Different cell types necessitate different settings. Check on the Neon transfection system web site for more information.

8. Act always as you are performing one further transfection (if you are supposed to transfect three times, resuspend the amount of cells needed for four transfections in 400 µl of resuspension buffer). This to avoid air bubble-related problems (detrimental for this kind of transfection), and to have spare cells in the case an extra transfection is required.

9. To obtain a good image of the mitochondrial network, mitochondrial fluorescence has to be really bright. This kind of fluorescence is rarely obtained by using antibodies for immunostaining. For this reason we recommend the transfection of a mitochondrial targeted fluorescent protein.

10. Starvation induces elongation of mitochondria [21]. Therefore, especially for long-time experiments, DMEM without phenol red is preferable to HBSS.

11. Slow rate of mito-PA-GFP signal diffusion implies a slower rate of mitochondrial fusion and vice-versa.

References

1. Corrado M, Scorrano L, Campello S (2012) Mitochondrial dynamics in cancer and neuro-degenerative and neuroinflammatory diseases. Int J Cell Biol 2012:729290. doi:10.1155/2012/729290

2. Detmer SA, Chan DC (2007) Functions and dysfunctions of mitochondrial dynamics. Nat Rev Mol Cell Biol 8:870–879. doi:10.1038/nrm2275

3. Youle RJ, van der Bliek AM (2012) Mitochondrial fission, fusion, and stress. Science 337:1062–1065. doi:10.1126/science.1219855

4. Campello S, Lacalle RA, Bettella M, Manes S, Scorrano L, Viola A (2006) Orchestration of lymphocyte chemotaxis by mitochondrial dynamics. J Exp Med 203:2879–2886. doi:10.1084/jem.20061877

5. Wang X, Su B, Lee HG, Li X, Perry G, Smith MA, Zhu X (2009) Impaired balance of mitochondrial fission and fusion in Alzheimer's disease. J Neurosci 29:9090–9103. doi:10.1523/JNEUROSCI.1357-09.2009

6. Ramonet D, Perier C, Recasens A, Dehay B, Bove J, Costa V, Scorrano L, Vila M (2013) Optic atrophy 1 mediates mitochondria remodeling and dopaminergic neurodegeneration linked to complex I deficiency. Cell Death Differ 20:77–85. doi:10.1038/cdd.2012.95

7. Itoh K, Nakamura K, Iijima M, Sesaki H (2013) Mitochondrial dynamics in neurodegeneration. Trends Cell Biol 23:64–71. doi:10.1016/j.tcb.2012.10.006

8. Shirendeb UP, Calkins MJ, Manczak M, Anekonda V, Dufour B, McBride JL, Mao P, Reddy PH (2012) Mutant huntingtin's interaction with mitochondrial protein Drp1 impairs mitochondrial biogenesis and causes defective axonal transport and synaptic degeneration in Huntington's disease. Hum Mol Genet 21:406–420. doi:10.1093/hmg/ddr475

9. Mitra K, Lippincott-Schwartz J (2010) Analysis of mitochondrial dynamics and functions using imaging approaches. Curr Protoc Cell Biol Chapter 4:Unit 4.25.1–Unit 4.2521. doi:10.1002/0471143030.cb0425s46

10. Shaner NC, Patterson GH, Davidson MW (2007) Advances in fluorescent protein technology. J Cell Sci 120:4247–4260. doi:10.1242/jcs.005801

11. Campbell RE, Tour O, Palmer AE, Steinbach PA, Baird GS, Zacharias DA, Tsien RY (2002) A monomeric red fluorescent protein. Proc Natl Acad Sci U S A 99:7877–7882. doi:10.1073/pnas.082243699

12. Strack RL, Strongin DE, Bhattacharyya D, Tao W, Berman A, Broxmeyer HE, Keenan RJ, Glick BS (2008) A noncytotoxic DsRed variant for whole-cell labeling. Nat Methods 5:955–957. doi:10.1038/nmeth.1264

13. Shaner NC, Steinbach PA, Tsien RY (2005) A guide to choosing fluorescent proteins. Nat Methods 2:905–909. doi:10.1038/nmeth819

14. Rizzo MA, Davidson MW, Piston DW (2009) Fluorescent protein tracking and detection: applications using fluorescent proteins in living cells. Cold Spring Harb Protoc 2009:pdb.top64. doi:10.1101/pdb.top64

15. Shitara H, Shimanuki M, Hayashi J, Yonekawa H (2010) Global imaging of mitochondrial morphology in tissues using transgenic mice expressing mitochondrially targeted enhanced green fluorescent protein. Exp Anim 59:99–103

16. Sterky FH, Lee S, Wibom R, Olson L, Larsson NG (2011) Impaired mitochondrial transport and Parkin-independent degeneration of respiratory chain-deficient dopamine neurons in vivo. Proc Natl Acad Sci U S A 108:12937–12942. doi:10.1073/pnas.1103295108

17. Wang Y, Pan Y, Price A, Martin LJ (2011) Generation and characterization of transgenic mice expressing mitochondrial targeted red fluorescent protein selectively in neurons: modeling mitochondriopathy in excitotoxicity and amyotrophic lateral sclerosis. Mol Neurodegener 6:75. doi:10.1186/1750-1326-6-75

18. Pickles S, Cadieux-Dion M, Alvarez JI, Lecuyer MA, Peyrard SL, Destroismaisons L, St-Onge L, Terouz S, Cossette P, Prat A, Vande Velde C (2013) Endo-MitoEGFP Mice: a novel transgenic mouse with fluorescently marked mitochondria in microvascular endothelial cells. PLoS One 8:e74603. doi:10.1371/journal.pone.0074603

19. Pham AH, McCaffery JM, Chan DC (2012) Mouse lines with photo-activatable mitochondria to study mitochondrial dynamics. Genesis 50:833–843. doi:10.1002/dvg.22050

20. Chudakov DM, Lukyanov S, Lukyanov KA (2007) Tracking intracellular protein movements using photoswitchable fluorescent proteins PS-CFP2 and Dendra2. Nat Protoc 2:2024–2032. doi:10.1038/nprot.2007.291

21. Gomes LC, Di Benedetto G, Scorrano L (2011) During autophagy mitochondria elongate, are spared from degradation and sustain cell viability. Nat Cell Biol 13:589–598. doi:10.1038/ncb2220

22. Karbowski M, Arnoult D, Chen H, Chan DC, Smith CL, Youle RJ (2004) Quantitation of mitochondrial dynamics by photolabeling of individual organelles shows that mitochondrial fusion is blocked during the Bax activation phase of apoptosis. J Cell Biol 164:493–499. doi:10.1083/jcb.200309082

Chapter 14

Analysis of Pro-apoptotic Protein Trafficking to and from Mitochondria

Ignacio Vega-Naredo, Teresa Cunha-Oliveira, Teresa L. Serafim, Vilma A. Sardao, and Paulo J. Oliveira

Abstract

Mitochondria play a key role in cell death and its regulation. The permeabilization of the outer mitochondrial membrane which is mainly controlled by proteins of the BCL-2 family, is a key event that can be directly induced by p53 and results in the release of pro-apoptotic factors to the cytosol, such as cytochrome c, second mitochondria derived activator of caspases/direct inhibitor-of-apoptosis (IAP) binding protein with low pI (SMAC/Diablo), Omi serine protease (Omi/HtrA2), apoptosis inducing factor (AIF), or endonuclease G (Endo-G). Hence, the determination of subcellular localization of these proteins is extremely important to predict cell fate and elucidate the specific mechanism of apoptosis. Here we describe the procedures that can be used to study the subcellular location of different pro-apoptotic proteins to be used in basic cell biology and toxicology studies.

Key words Mitochondria, Pro-apoptotic proteins, Cell fractions, Immunoblot, Immunoprecipitation, Immunocytochemistry

1 Introduction

Mitochondria are cellular organelles with important functions in cell life and death, acting as cellular powerhouses by producing energy to maintain cellular activity. However, mitochondria are also important checkpoints for cell fate decisions, playing a crucial role in programmed cell death (PCD) pathways. The permeabilization of the outer mitochondrial membrane (OMM) is a fundamental step in several tightly regulated pathways of cell death, allowing the release of proteins that are usually only present in the intermembrane space, and signaling cell death programs [1]. One of the best characterized types of cell death is apoptosis. Apoptotic signals may originate outside the cell (extrinsic pathway) or from any intracellular compartment (intrinsic pathway), constituting two distinct yet complementary apoptotic mechanisms. Both intrinsic and extrinsic stimuli may lead to OMM permeabilization.

Carlos M. Palmeira and Anabela P. Rolo (eds.), *Mitochondrial Regulation*, Methods in Molecular Biology, vol. 1241, DOI 10.1007/978-1-4939-1875-1_14, © Springer Science+Business Media New York 2015

The OMM is selectively permeable to solutes, and its permeability and integrity are regulated by proteins of the BCL-2 family [2]. This family includes pro- and anti-apoptotic members, depending on the presence of different BCL-2 homology (BH) domains in their structure, conferring different functions [3]. Anti-apoptotic proteins, such as BCL-2 and BCL-xL have four different BH domains—BH1234. Pro-apoptotic proteins such as BAX and BAK are also multi-domain proteins with three different BH domains—BH123. BH3-only proteins are pro-apoptotic proteins with only one BH domain and can exert their pro-apoptotic function either by facilitating or by activating BH123 proteins, which then initiate OMM permeabilization. Facilitators or de-repressors, such as BAD, interact with BH1234 proteins, dissociating them from sequestered pro-apoptotic proteins, which become free to promote OMM permeabilization. The activators, such as tBID (which results from the cleavage of BID by caspase-8), directly activate BH123 proteins, either by stimulating the translocation of Bad to the OMM, or by interacting with BAK. OMM permeabilization may occur through a Bax/Bak-mediated mechanism [4] or by the opening of the mitochondrial permeability transition pore (MPTP) in the inner mitochondrial membrane [5]. In the latter mechanism, opening of the MPTP can lead to mechanical rupture of the OMM to which pro-apoptotic protein release follows or instead to the recruitment of pro-apoptotic proteins to the OMM by causing mitochondrial depolarization [6].

P53 is a redox-sensitive transcription factor with a broad range of actions, some of them related to survival and cell death [7]. In general, the tumor suppressor p53 exerts important roles in cell cycle progression and cell death coordinating multiple options for cellular response to genotoxic stress. p53 inhibits replication of the genome by blocking cell cycle progression at a G1/S check point in response to DNA damage [8]. In unstressed cells, p53 levels are low but its expression increase following stress signals acting through both transcription-dependent and -independent mechanisms to coordinate the appropriated cellular responses [9]. p53 activity depends on its ability to activate or repress gene transcription. Thus, p53 oscillates between latent and active sequence-specific DNA binding conformations and is differentially activated through posttranslational modifications including phosphorylation, acetylation and ubiquitination. On the other side, nonsequence-specific DNA binding may mediate other p53 actions [10]. In addition, p53 is also involved in mitochondrial-dependent cell death, collaborating in the execution of the apoptotic pathway. In this context, p53 undergoes a nuclear–cytoplasm–mitochondria trafficking and western blotting and immunocytochemistry are usually performed to detect the presence of p53 in nuclear, cytoplasmic, and mitochondrial extracts isolated as described in this paper. The analysis of immunoreactive bands in nuclear fractions indicates whether nuclear translocation is occurring. The nuclear localization is critical for its transcriptional activity by activating genes that arrest cell growth and repair DNA damage. To further confirm its transactivation function, evaluation of the expression of the transcription target genes of p53 such as PUMA, NOXA, BAX, BID, and DRAM is often useful [11]. As described above, the activity of the p53 gene product is regulated by posttranslational modifications. These modifications of p53 affect its stability and can be a potential mechanism to select the target genes conferring differential binding affinity to the response elements. For example, acetylation influences p53 activity enhancing its transcriptional activity and can be detected using different acetyl-p53 antibodies available. Acetylation of p53 at carboxyl-terminal lysine residues enhances its transcriptional activity associated with cell cycle arrest and apoptosis. However, p53 acetylation at Lys-320/Lys-373/ Lys-382 is also required for transcription-independent functions involving BAX activation [12]. The phosphorylation of p53 at Ser15 and Ser20 following DNA damage can also be detected by immunoblotting to evaluate its transactivation function since this phosphorylation promotes p53 activation and stabilization reducing the interaction between p53 and its negative regulator, the oncoprotein murine double minute 2 (MDM2) [13]. In addition, phosphorylation at Ser392 influences transcriptional activation of p53 and regulates the oncogenic function of p53 [14, 15]. MDM2 inhibits cytoplasmic retention of p53 by targeting it for ubiquitination and proteasomal degradation [16]. Some reports suggest not only that the cytoplasmic retention of p53 is able to repress autophagy [17, 18] but also that this retention can be mediated by acetylation since p53 acetylation reduces its ubiquitination status [19]. Although the precise molecular mechanisms behind this remain unclear, p53 deacetylases may be upregulated to mediate ubiquitination and degradation of p53 [20]. Because of this, the immunoblot analysis of cytoplasmic extracts can reveal whether p53 is accumulated in cytoplasm and can help to reveal the function of p53 in a specific context. p53 can promote cell death independently of transcription by two different mechanisms, each of which assigned to a specific localization of the protein: cytosol or mitochondrial. Both modes of action converge in the permeabilization of the OMM via activation of the pro-apoptotic proteins BAX or BAK. In fact, cytosolic p53 can directly activate BAX and thereby induce apoptosis [21]. On the other hand, in response to a broad spectrum of apoptotic stimuli, a fraction of p53 translocates to mitochondria and triggers a direct mitochondrial p53 death program [22]. For this, p53 physically interacts with the BCL-2 family member proteins BCL-xL and BCL-2 antagonizing their anti-apoptotic function and inducing OMM permeabilization [23]. Furthermore, mitochondrial p53 directly promotes the pro-apoptotic activities of BAX and directly induces

BAK oligomerization [24]. p53 also interacts with the antioxidant enzyme superoxide dismutase 2 (SOD2) leading to a reduction of its superoxide scavenging activity, and a subsequent decrease of mitochondrial membrane potential which contributes to the induction of pro-apoptotic mitochondrial alterations [25]. Surprisingly, p53 is released from mitochondria mediating a retrograde signaling pathway to the nucleus [26]. Therefore, the study of the subcellular localization of p53, its posttranslational modifications and the levels of p53-related/-targeted proteins are crucial to examine the p53 pathway and elucidate its particular role.

As described above, OMM permeabilization resulting from stress stimuli can result in simultaneous release of pro-apoptotic factors that are normally limited to the mitochondrial intermembrane or intercristae space, including cytochrome c, apoptosis-inducing factor (AIF), endonuclease G (endoG), Smac/Diablo, and Omi/HtrA2 [27]. Cytochrome c was first identified as being involved in mitochondrial bioenergetics, essential for the ATP production by oxidative phosphorylation, and later to apoptosis. Following the release of cytochrome c from the intermembrane space, it binds to Apoptotic protease activating factor 1 (APAF-1) and dATP [28], forming the apoptosome complex and initiating cell death with the activation of caspase-9 and consequent triggering of caspase cascade [29]. The second mitochondrial-derived activator of caspase (Smac), also known as direct IAP-binding protein with a low pI (Diablo) [29], resides in mitochondria in a mature form, and is released during apoptosis, which promotes caspase-dependent apoptosis by controlling the activity of inhibitor of apoptosis protein (IAP). Omi is a serine protease, which can be released from mitochondrial intermembrane space into cytoplasm, upon apoptotic insult. Omi cleaves both IAPs and cytoskeletal proteins, contributing to apoptosis in caspase-dependent and independent manners [30].

Independently of caspase activation, cell death can occur with the release of endoG and AIF. EndoG is also localized in intermembrane space bound to the inner membrane. Due to mitochondrial membrane potential loss, the mature EndoG is translocated to the nucleus and initiates oligonucleosomal DNA fragmentation. EndoG can regulate several mitochondrial enzymes expression, such as complex I (ND1 and ND2), complex IV (COX2), and complex V (ATPase6) [30] (Fig. 1). The AIF is a flavoprotein harboring NADH oxidase activity. Initially, the AIF was identified as a pro-apoptotic protein, inducing a type of programmed cell death independently of caspases activation. When AIF is added to purified nuclei extracts in a cell free system, chromatin condensation and large-scale DNA fragmentation to ~50 kbp fragments occurs [31]. Those effects were observed in intact cells after diverse apoptotic stimuli and also in models of retinal degeneration, brain damage induced by hypoglycemia or ischemia, or myocardial infarction [32].

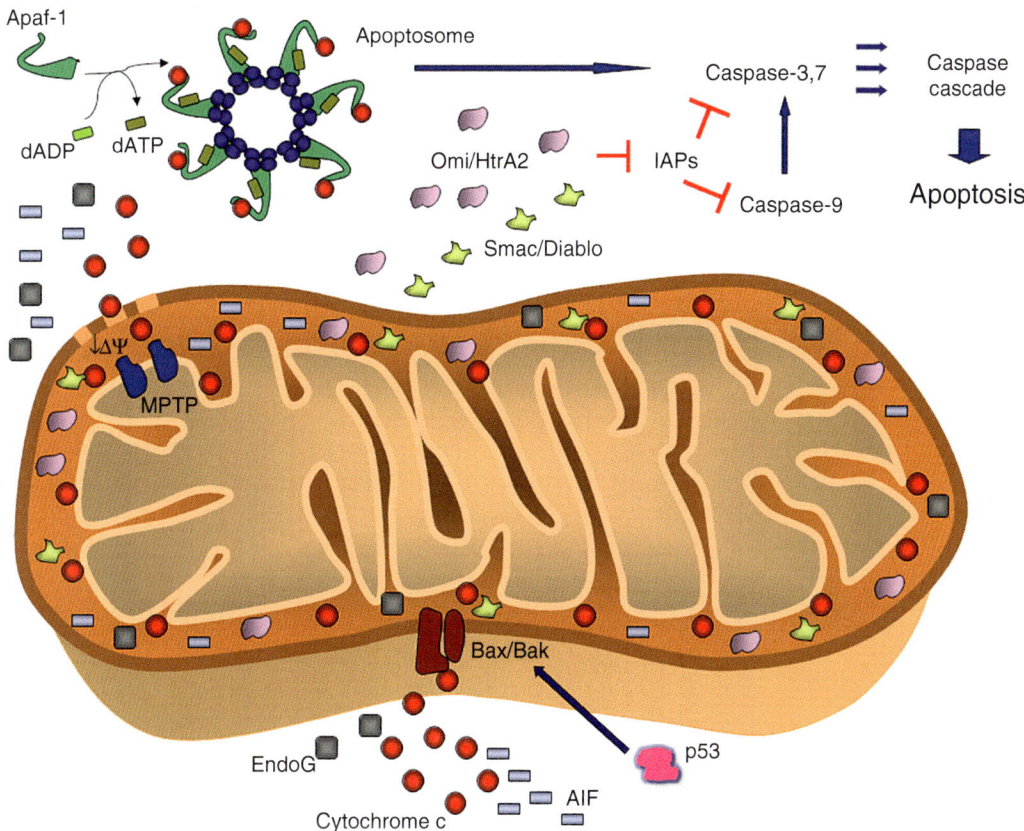

Fig. 1 Intrinsic apoptotic pathway. The intrinsic or mitochondrial apoptotic pathway, initiates with mitochondrial membrane permeabilization, which may result from p53 signaling, with formation of different pores in mitochondrial membrane, such as Bax/Bak or mitochondrial permeability transition pore (MPTP), in the latter case by leading to OMM rupture or by triggering pro-apoptotic protein translocation to mitochondria following mitochondrial depolarization. Consequently, the release of apoptotic factors to cytosol, including cytochrome c, second mitochondria derived activator of caspases/direct inhibitor-of-apoptosis binding protein with low p/ (SMAC/Diablo), Omi serine protease (Omi/HrtA2), endonuclease G (endoG), and apoptosis-inducing factor (AIF), occurs. The released cytochrome c reacts with deoxy ATP (dATP) and apoptotic protease activating factor 1 (APAF-1), leading to apoptosome formation. The apoptosome recruits procaspase-9, activating the initiator caspase-9 and caspase cascade begins

The AIF protein is encoded by a nuclear gene, synthesized in the cytoplasm and transported to mitochondria by the general import pathway, due to the mitochondrial localization sequence (MLS) in the precursor protein. Once in the mitochondrial intermembrane space, the MLS is proteolytically cleaved, the protein refolds and incorporates flavin adenine dinucleotide (FAD) cofactor, generating the mature protein. The oxidoreductase activity, dependent of the presence of prosthetic group, is not critical for the apoptogenic effect of AIF but the structural parts of the oxidoreductase domain are necessary to the DNA biding [31].

A decrease in AIF enzymatic activity or a decrease in AIF expression results in decreased oxidative phosphorylation and increased free radicals generation [33]. On the other hand and depending on cell type and on the cell death stimuli, AIF protein is involved in programmed cell death. The basic mechanism of AIF-induced cell death consists in AIF release from the mitochondrial intermembrane space to the cytosol and then on its translocation to the nucleus. Once in the nucleus, AIF binds to DNA induces chromatin condensation and large-scale DNA fragmentation. The nuclear apoptosis induced by AIF requires a direct interaction of AIF with DNA [34]. In the cytosol, AIF also promotes a decrease in mitochondrial $\Delta\Psi$ and release of cytochrome c and further AIF release from mitochondria, promoting a positive feedback amplification loop [31, 32]. In order to be released from the mitochondrial intermembrane space, AIF must be cleaved in a specific region in order to release the protein binding to the mitochondrial inner membrane. This cleavage is performed by proteases such as calpains and cathepsins that may have access to the mitochondrial intermembrane space during apoptotic stimuli and require the presence of calcium. The truncated AIF (tAIF) is then free to execute its caspase-independent apoptotic action. An interesting aspect is the fact that binding of AIF to DNA induces large-scale DNA fragmentation, but AIF itself does not possess DNAse activity. Thus, it has been proposed that the DNA-degrading capacity of AIF could be due to the recruitment of downstream nucleases, such as cyclophilin A (CypA) [35, 36].

In this chapter, we describe the material and the methods followed by us and other authors to study the pro-apoptotic protein trafficking to and from mitochondria: immunoblotting performed with different cellular fractions and immunocytochemistry using the antibodies described in Table 1. Immunocytochemical approaches can be used to study subcellular localization of pro-apoptotic factors described in this chapter. To evaluate their mitochondrial localization, co-localization assays using antibodies constitute an effective alternative (Fig. 2). One option is to develop an immunocytochemistry using a secondary antibody conjugated with a fluorochrome such as FITC and to label mitochondria using mitochondria-selective probes such as MitoTracker Red. Another option consists in a double immunofluorescence assay using a mixture of two primary antibodies, for example one against the mitochondrial marker TOM20 and the other against the desired pro-apoptotic protein, using their respective secondary antibodies which have to be raised in different species and conjugated with two different fluorochromes (e.g., FITC-conjugated against rabbit and Texas Red-conjugated against mouse).

Table 1
List of antibodies for western blotting (WB) and immunocytochemistry (ICC)

Antibodies	MW (kDa)	Company (cat. no.)	Species cross-reactivity	Isotype	Applications (recommended dilution)
p53	53	Cell Signaling (2524)	H, M, R, Mk	Mouse IgG1	WB (1:1,000); ICC (1:250)
p53	53	Santa Cruz (sc-6243)	H, M, R	Rabbit IgG	WB (1:500); ICC (1:50)
Phospho-p53 (Ser15)	53	BioVision (3515)	H, M, R	Rabbit IgG	WB (4 µg/ml)
Phospho-p53 (Ser15)	53	Cell Signaling (9298)	H, M, R, Mk	Rabbit IgG	WB (1:1,000); ICC (1:250)
Phospho-p53 (Ser329)	53	Cell Signaling (9281)	H, M	Rabbit IgG	WB (1:1,000)
Acetyl-p53 (Lys379)	53	Cell Signaling (2570)	H, M	Rabbit IgG	WB (1:1,000)
Puma	23	Cell Signaling (4976)	H	Rabbit IgG	WB (1:1,000)
Puma	23	Cell Signaling (7467)	M, R	Rabbit IgG	WB (1:1,000)
Noxa	15	Santa Cruz (sc-56169)	H, M	Mouse IgG1	WB (1:500)
Bax	20	Cell Signaling (2772)	H, M, R, Mk	Rabbit IgG	WB (1:1,000)
Bak	25	Cell Signaling (3814)	H, M, R, Mk	Rabbit IgG	WB (1:1,000)
Bid	22 (15)	Cell Signaling (2002)	H	Rabbit IgG	WB (1:1,000)
Bid	22	Cell Signaling (2003)	M	Rabbit IgG	WB (1:1,000)
DRAM	33	Rockland (600-401-A70)	H, M, R	Rabbit IgG	WB (4 µg/ml); ICC
MDM2	90 (60)	Santa Cruz (sc-965)	H, M, R	Mouse IgG1	WB (1:500); ICC (1:50)
SOD2	25	Santa Cruz (sc-18504)	H, M, R	Goat IgG	WB (1:500); ICC (1:50)
Ubiquitin		Cell Signaling (3933)	H, M, R, Mk	Rabbit IgG	WB (1:1,000)
Bcl-2	26	Cell Signaling (2870)	H, M, R, Mk	Rabbit IgG	WB (1:1,000)

(continued)

Table 1
(continued)

Antibodies	MW (kDa)	Company (cat. no.)	Species cross-reactivity	Isotype	Applications (recommended dilution)
Bcl-xL	30	Cell Signaling (2764)	H, M, R, Mk	Rabbit IgG	WB (1:1,000); ICC (1:200)
Tom20	20	Santa Cruz (sc-11415)	H, M, R	Rabbit IgG	WB (1:500); ICC (1:100)
Cytochrome c	12	Abcam (ab13575)	M, R, H, P, Hr, Ff	Mouse IgG	WB (1:500); ICC (1:200)
Cytochrome c	12	Mitosciences (37BA11)	M, R, C, H, Ce	Mouse IgG	WB (0.5 μg/ml); ICC (1 μg/ml)
Smac/Diablo	21	Cell Signaling (2954)	H, Mk	Mouse IgG	WB (1:1,000); ICC (1:100)
Apaf-1	130	Millipore (AB16503)	H, R, M	Rabbit IgG	WB (1:1,000)
Omi	48	Abcam (ab33041)	M, H	Mouse IgG	ELISA; WB (1:1,000)
AIF	57	Santa Cruz (sc-13116)	H, R, M	Mouse gG2b	WB (1:1,000); ICC (1:100)
Anti-rabbit IgG-AP		Santa Cruz (sc-2007)		Goat IgG	WB (1:5,000)
Anti-goat IgG-AP		Santa Cruz (sc-2022)		Donkey IgG	WB (1:5,000)
Anti-mouse IgG1-AP		Santa Cruz (sc-2066)		Goat IgG	WB (1:5,000)
Anti-rabbit IgG-FITC		Santa Cruz (sc-2012)		Goat IgG	ICC (1:400)
Anti-mouse IgG-TR		Santa Cruz (sc-2781)		Goat IgG	ICC (1:400)

Ce Caenorhabditis elegans, *C* cow, *Ff* fruit fly, *Hr* horse, *H* human, *Mk* monkey, *M* mouse, *P* pigeon, *R* rat

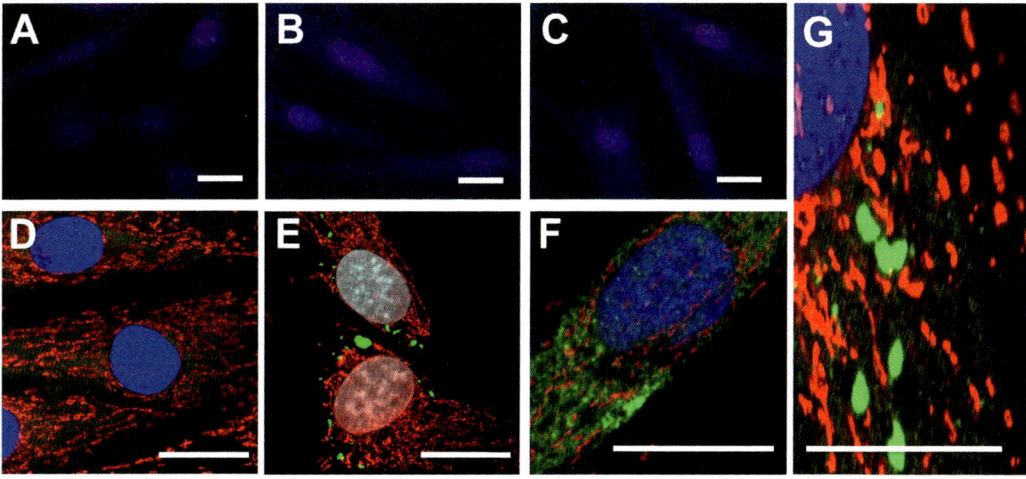

Fig. 2 Representative micrographs of fluorescence microscopy of immunocytochemistry-labeled apoptotic proteins. Images show H9c2 cardiomyoblasts treated with the anticancer agent Doxorubicin (DOX) and imaged for p53 (**a–c**) and Bax (**d–g**). Panels **a–c** show H9c2 cells treated with vehicle (**a**), 0.5 μM (**b**) and 1 μM (**c**) DOX for 24 h. After treatment, H9c2 cells were fixed with ice cold methanol (which removed nuclear bound DOX), labeled with an antibody against p53. Epifluorescence microscopy images in panels **a–c** illustrate increased p53 nuclear labeling (in *blue*) after H9c2 cells treatment with DOX. Cells were observed by epifluorescence microscopy using a Nikon Eclipse TE2000U microscope equipped with a 40× Plan Fluor 1.3 NA oil immersion DIC objective and images were processed using Metamorph software (Universal Imaging, Downingtoen, PA). Panels **d–g** show confocal microscopy images of H9c2 cells labeled with Hoechst 33342 (nucleus, *blue* and *pink*), Mitotracker Red (mitochondria, *red*) and Bax (*green*). Confocal microscopy of H9c2 cells treated with 0 (**d**), 0.5 μM (**e, g**) and 1 μM (**f**) DOX for 24 h. After treatment, cells were fixed in paraformaldehyde (in order to maintain the integrity of the mitochondrial network) and subsequently immunolabeled with an antibody against Bax. Increased immunolabeling for Bax is observed after DOX treatment, especially forming large clusters in the cytosol of treated cells. This is clearly visible in panel **g**, obtained with a higher magnification. Images in panels **d–g** were obtained by using a Nikon C-1 laser scanning confocal microscope equipped with a 60× Plan Apo 1.4 NA oil immersion DIC objective. Images were captured using the Nikon EZ-C1 software (version 2.01). *White bar* in all panels represents 20 μm

2   Materials

2.1 Nuclear, Mitochondrial, and Cytosolic Fraction Isolation from Cultured Cells

1. PBS: 137.93 mM NaCl, 2.67 mM KCl, 8.06 mM Na_2HPO_4, 1.47 mM KH_2PO_4, pH 7.4.

2. Buffer A: 250 mM Sucrose, 20 mM Hepes, 10 mM KCl, 1.5 mM $MgCl_2$, 0.1 mM EDTA, 1 mM EGTA, pH 7.5 (adjusted with KOH).

3. Buffer B: 250 mM Sucrose, 10 mM $MgCl_2$.

4. Buffer C: 350 mM Sucrose, 0.5 mM $MgCl_2$.

5. Nuclear buffer: 5 mM HEPES, 1.5 mM $MgCl_2$, 0.2 mM EDTA, 0.5 mM DTT, 26 % glycerol (v/v), pH 7.9.

2.2 Western Blotting

1. Protein quantification: Bradford (B6916; Sigma, St. Quentin Fallavier, France) or bicinchoninic acid (BCA) (23227; Thermo Scientific, Rockford, IL, USA) assays.

2. Laemmli buffer: 62.5 mM Tris–HCl at pH 6.8, 25 % glycerol, 2 % SDS, 0.01 % bromophenol blue. Add 5 % β-mercaptoethanol prior to use (*see* **Note 1**).

3. Mini-PROTEAN system with Mini Trans-Blot module, gel cassettes and casting stand, short and spacer plates, combs, external power supply (Bio-Rad, Hercules, CA, USA) and rollers or shakers for incubations.

4. Prestained protein standard (161-0374; Bio-Rad), polyvinylidene difluoride (PVDF) membranes (Millipore, Bedford, MA) and Ponceau S staining solution: 0.1 % (w/v) in 5 % acetic acid.

5. Buffers: Running buffer (25 mM Tris, 192 mM glycine, 0.1 % SDS, pH 8.3); Transfer buffer (25 mM Tris, 192 mM glycine, 20 % methanol) and TBS-T (10 mM Tris–HCl at pH 8.0, 150 mM NaCl, 0.1 % Tween-20).

6. ECF substrate (RPN5785; GE Healthcare, Munich, Germany) and a detection system like VersaDoc (Bio-Rad).

2.3 Immunoprecipitation

1. Protein G PLUS-Agarose (sc-2002, Santa Cruz Biotechnology, Santa Cruz, CA, USA).

2. IP buffer I (50 mM Tris–HCl, pH 7.4, 150 mM NaCl, 1 % Na-deoxycholate, 1 % NP-40, 1 mM Na_3VO_4).

3. IP buffer II (50 mM Tris–HCl, pH 7.4, 75 mM NaCl, 0.1 % Na-deoxycholate, 0.1 % NP-40, 1 mM Na_3VO_4).

4. IP buffer III (50 mM Tris–HCl, pH 7.4, 0.05 % Na-deoxycholate, 0.05 % NP-40, 1 mM Na_3VO_4).

5. Laemmli buffer: 62.5 mM Tris–HCl at pH 6.8, 25 % glycerol, 2 % SDS, 0.01 % bromophenol blue. Add 5 % β-mercaptoethanol prior to use.

2.4 Immunocytochemistry

1. Sterile glass coverslips, 6-well plates, lancets, forceps, microscope slides and a dark humidified chamber.

2. Culture medium and mitochondrial fluorescent probes: MitoTracker Red CMXRos (M7512; Invitrogen, Paisley, UK) or MitoTracker Green FM (M7514; Invitrogen).

3. PBS: 137 mM NaCl in 10 mM phosphate buffer, pH 7.4.

4. 4 % Formaldehyde in PBS, 0.2 % Triton X-100 in PBS and 1 % BSA in PBS.

5. Prolong Gold antifade medium with DAPI (P36935; Invitrogen) or without DAPI (P36934; Invitrogen) and nail polish.

3 Methods

3.1 Nuclear, Mitochondrial, and Cytosolic Fraction Isolation from Cells in Culture

1. Supplement buffer A with 1 μg/ml of leupeptin, antipain, chymostatin, and pepstatin A, 1 mM of DTT and 100 μM of PMSF. Supplement buffer B and C with 1 mM of DTT and 100 μM of PMSF. Supplement the nuclear buffer with 300 mM of NaCl (high salt helps lyse nuclear membranes and forces DNA into solution).

2. Grow cells on cell culture dishes in an appropriated cell culture medium and perform the desired treatment (*see* **Note 2**).

3. After treatment, aspirate or collect (if you are interested in the floating cells) the incubation media. Rinse the adherent cells with 5 ml of PBS and waste it.

4. Harvest adherent cells with 3–5 ml of trypsin. Inhibit the trypsin with 3–5 ml of growth media containing FBS (*see* **Note 3**).

5. Centrifuge between 300 and $400 \times g$ for 5 min at 4 °C in order to collect all cells without damaging membrane integrity.

6. Aspirate the supernatant and rinse the cell pellet with 2 ml of PBS. Perform again the centrifugation step, discard the supernatant and resuspend the pellet in 1 ml of complete buffer A. Incubate for 15 min on ice.

7. Homogenize cells in a Potter-Elvehjem homogenizer with a Teflon pestle (30–40 strokes), or alternatively pass cell suspension through a 25 G needle ten times using a 1 ml syringe. This procedure should also be performed on ice.

8. Centrifuge the cellular suspension at $720 \times g$ for 5 min at 4 °C. Remove the supernatant (containing mitochondrial and cytosolic fractions) and keep it on ice.

9. Resuspend the nuclear pellet again in 1 ml of complete buffer A. Homogenize the pellet again in a Potter–Elvehjem homogenizer with a Teflon pestle, or alternatively pass through a 25 G needle. Centrifuge again at $720 \times g$ for 10 min at 4 °C. Discard the supernatant and resuspend the pellet in 0.5 ml of complete buffer B and pour the nuclear suspension on the complete buffer C. Centrifuge at $1,430 \times g$ for 5 min at 4 °C. Remove the supernatant and discard it. Resuspend the nuclear pellet in 50 μl of nuclear buffer supplemented with 300 mM of NaCl. Homogenize the nuclear pellet passing nuclear suspension through a 27 G needle ten times using a 0.5 ml syringe. Store at –80 °C until further analysis.

10. Centrifuge the supernatant collected in **step 8** at $14,000 \times g$ for 10 min at 4 °C. Resuspend the pellet (mitochondrial fraction) in 50 μl of buffer 1 and store at –80 °C until further analysis.

11. Collect the supernatant (containing the cytosolic fraction) and centrifuge in an ultracentrifuge at $100,000 \times g$ for 30 min at

4 °C. Discard the pellet (containing the membrane fractions) and concentrate the supernatant (containing the cytosolic fraction) by lyophilization or by tangential flow filtration. Store at −80 °C until further analysis be performed (*see* **Note 4**).

In Fig. 3, a flowchart outlining the isolation procedure is presented

3.2 Western Blotting

3.2.1 SDS-PAGE

1. Determine protein content of all fractions by standard procedures such as the Bradford or the BCA method using 5 μl of aliquot.

2. Mix the samples with the appropriated volume of Laemmli buffer (1:1).

3. Prepare the Sodium Dodecyl Sulfate (SDS) 10 % polyacrylamide resolving gel (*see* **Note 5**) by mixing 2.5 ml Tris–HCl 1.5 M pH 8.8, 0.1 ml SDS 10 %, 50 μl ammonium persulfate 10 %, 2.5 ml acrylamide/bisacrylamide (29:1) 40 %, 4.9 ml of distilled water (for a 0.75–1.0 mm thick gel). Add 5 μl of TEMED to initiate polymerization. Cast gel within an assembled gel cassette allowing space for stacking gel, gently overlay with water and wait until polymerization.

4. Prepare the stacking gel (4 % polyacrylamide) by mixing 1.25 ml Tris–HCl 0.5 M, pH 6.8, 50 μl SDS 10 %, 25 μl ammonium persulfate 10 %, 0.5 ml acrylamide–bisacrylamide (29:1) 40 %, 3.2 ml of distilled water, and 5 μl of TEMED. Insert a gel comb immediately and wait until polymerization.

5. Denature the samples by boiling at 95–100 °C for 5 min.

6. For polyacrylamide gel electrophoresis, we normally use Mini-PROTEAN systems from Bio-Rad, but other alternative systems can be used. After complete polymerization, place the gel into the electrophoretic chamber with running buffer, load the volume of sample corresponding to 5–25 μg of protein in each individual well (*see* **Note 6**), as well as 6 μl of prestained protein standard (e.g., Precision Plus Protein Dual Color Standards from Bio-Rad) into one of the other lanes.

7. Run electrophoresis at constant voltage (100–120 V). Running can be monitored by observing migration of prestained protein standards and bromophenol blue front. Stop running when the bromophenol blue band leaves the lower end of the gel. Keep gels in running buffer until ready to transfer.

3.2.2 Wet/Tank Electrophoretic Transfer

1. Activate polyvinylidene difluoride (PVDF) membrane in methanol for 15 s. Transfer the membrane from methanol to transfer buffer and incubate on shaker for at least 5 min. Soak also pads, filter papers, and the gel in transfer buffer before use.

2. Assemble the transfer sandwich in a shallow tray filled with transfer buffer as follows: Black side of the sandwich (cathode),

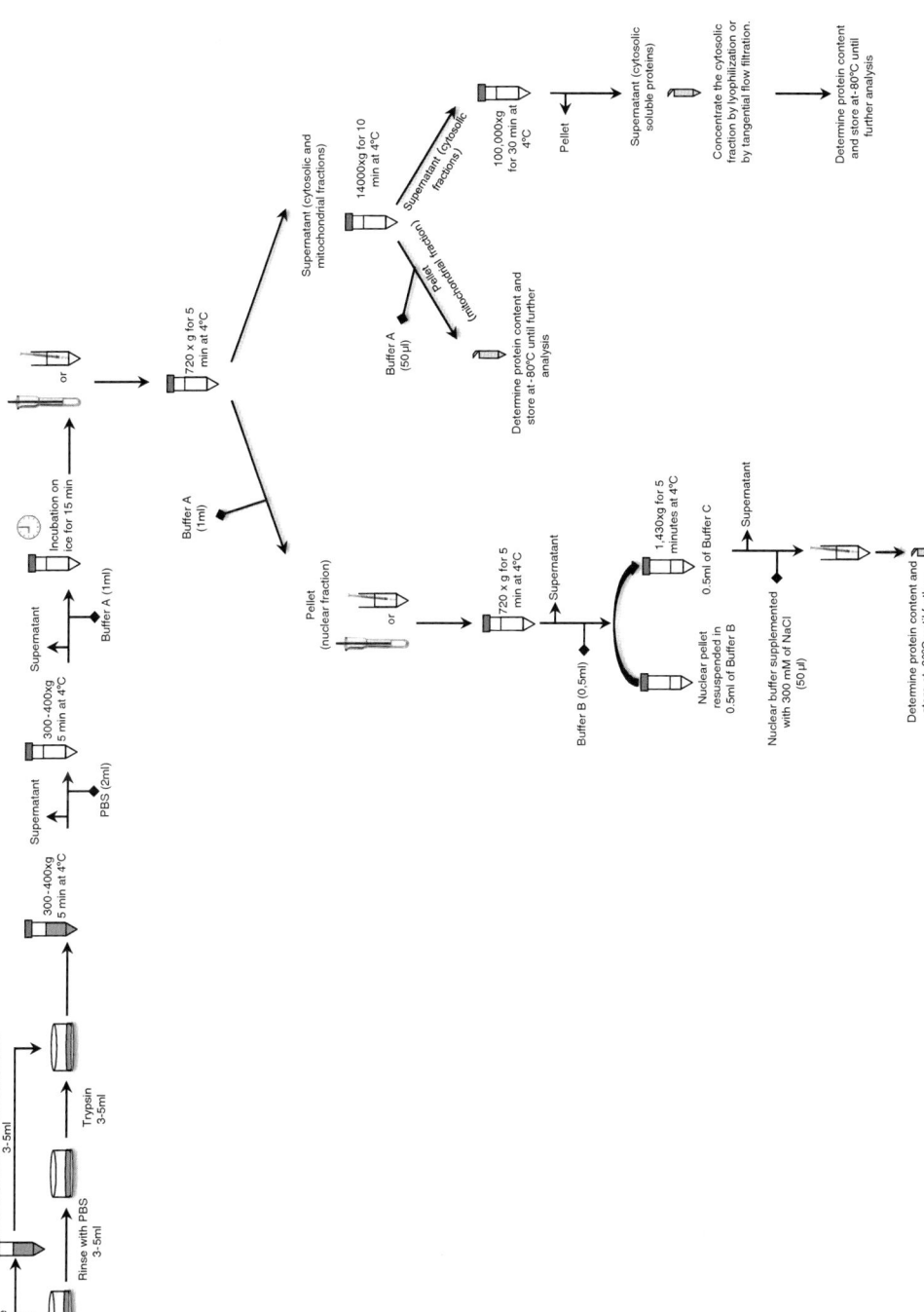

Fig. 3 Flowchart outlining the isolation of nuclear, mitochondrial, and cytosolic fractions from cells in culture. All the material was kept on ice during all the procedures and all the centrifugations were performed at 4 °C

soaked pads, filter paper, gel, PVDF membrane, filter paper, soaked pad, and white side of the sandwich (anode). Avoid bubbles formation between gel and PVDF membrane. Fill tank apparatus with transfer buffer and run at 350 mA for 90 min with an ice pack or another cooling system (*see* **Note 7**).

3. Stain the membrane with Ponceau S for 5 min to check protein transfer. Finally, transfer the membrane to TBS-T to rinse Ponceau staining.

3.2.3 Enzymatic Immunodetection

1. Block the membrane to reduce nonspecific binding with 5 % skim milk in TBS-T overnight at 4 °C or alternatively for 2 h at room temperature (*see* **Notes 8** and **9**).

2. Incubate the membrane with the primary antibody at the recommended dilution (*see* Table 1) in 1 % skim milk in TBT-T for 4 h at 4 °C or overnight.

3. Wash three times with TBS-T for 5 min at room temperature.

4. Incubate the membrane with the corresponding alkaline phosphatase conjugated secondary antibody (1:5,000) in 1 % skim milk in TBT-T for 2 h at 4 °C.

5. Wash three times with TBS-T for 20 min at room temperature.

6. Detect the immunoconjugates using the western blotting ECF substrate (or similar) according to the manufacturer's protocol. Use fluorescence scanning equipment such as VersaDoc (Bio-Rad) to develop the image and analyze it using the software QuantityOne (Bio-Rad) or Image J (or similar). Please notice that local vs. global background subtraction must be chosen since this will affect the final results.

3.3 Immuno-precipitation

P53 ubiquitination and the physical interaction of p53 with some related proteins such as Bcl-2, Bcl-xL, Bax, and SOD-2 can be evaluated by p53 immunoprecipitation from homogenates and subsequent immunoblot analysis with antibodies against these related proteins.

1. Incubate a volume of extracts corresponding to equal quantity of protein (e.g., 300 μg of protein) with 5 μl mouse anti-p53 (2524; Cell Signaling Technology) for 1 h at 4 °C under orbital shaking conditions.

2. Add 20 μl of protein G PLUS-Agarose and incubate overnight at 4 °C under orbital shaking conditions.

3. Centrifuge 1 min at $16,000 \times g$ and discard the supernatant fraction.

4. Add 1 ml of IP buffer I to the pellet and mix the sample on an oscillatory shaker for 20 min at 4 °C.

5. Centrifuge for 1 min at $16,000 \times g$ and discard the supernatant fraction.

6. Repeat with IP buffer II.

7. Repeat with IP buffer III.

8. Centrifuge 1 min at $16,000 \times g$ and discard the supernatant fraction.

9. Add 20 μl of Laemmli buffer and boil for 5 min.

10. Centrifuge again 1 min at $16,000 \times g$ to pellet the agarose beads.

11. Transfer the supernatant fraction to a new tube, subject it to SDS-PAGE and immunoblot analysis as previously described (in Subheading 3.2) using primary antibodies against p53 (sc-6243; Santa Cruz Biotechnology), BCL-2, BCL-xL, BAX, SOD-2 and/or ubiquitin (check Table 1 for references and recommended dilutions).

3.4 Immunocyto-chemistry

3.4.1 Cell Seeding for Morphological Studies

1. Seed cells on glass coverslips in 6-well plates and wait for 24 h for cell adhesion at 37 °C in a 5 % CO_2 atmosphere.

3.4.2 Staining with Mitochondrial Specific Fluorescent Dye (Optional)

1. Remove culture medium and incubate cells with 125 nM MitoTracker Red CMXRos (M7512; Invitrogen) in culture medium for 20 min at 37 °C. Alternatively, 0.5 μM green-fluorescing MitoTracker Green FM (M7514; Invitrogen) for 30 min at 37 °C can be used to label mitochondria (*see* **Note 10**).

2. Replace staining solution with fresh pre-warmed media and subject cells to subsequent processing steps.

3.4.3 Fixation and Permeabilization

1. Remove the incubation media and fix cells in 4 % formaldehyde in PBS for 15 min at 37 °C.

2. Rinse three times with PBS for 5 min each.

3. Permeabilize cells with 0.2 % Triton X-100 in PBS for 10 min.

4. Rinse three times with PBS for 5 min each.

3.4.4 Immunolabeling

1. Block to reduce nonspecific binding with 1 % BSA in PBS for 1 h at 4 °C (*see* **Note 11**).

2. Distribute 100–150 μl of primary antibody (or mixture of antibodies) at 1:250 in PBS, 1 % BSA on each coverslip and incubate 90 min in a humidified chamber at room temperature.

3. Remove the primary antibody (it can be stored in 0.05 % sodium azide) and rinse three times with PBS for 5 min.

4. Incubate with the corresponding fluorescence-conjugated secondary antibody (or mixture of antibodies) at 1:400 in PBS, 1 % BSA for 1 h in a dark humidified chamber at 37 °C.

5. Remove the secondary antibody and rinse with PBS three times for 5 min each with PBS.

3.4.5 *Mounting Coverslips*

1. Mount coverslips on glass slides using Prolong Gold antifade medium (P36934; Invitrogen). If nuclear labeling is desired, use Prolong Gold antifade medium with DAPI (P36935; Invitrogen).

2. Allow to dry overnight, seal with nail polish, and store at –80 °C until analysis under a confocal microscope (LSM 510Meta; Zeiss).

4   Notes

1. Alternatively to the use of β-mercaptoethanol in Western Blot sample preparation, dithiothreitol (DTT) can be used in the same proportion.

2. In order to have enough protein for the different cellular fractions, it is often necessary to start with large amount of cells, especially for large volume cells. Optimization process for cell density is required.

3. If the incubation medium is saved in order to collect floating cells, it can be used to inhibit the trypsin.

4. It is advisable to test the purity of each sample using specific antibodies for each fraction. For example, one strategy involves cross-labeling samples from different fractions with antibodies against a particular histone (nuclear marker), the voltage-dependent anion channel (VDAC) or TOM20 (mitochondrial markers), and glyceraldehyde 3-phosphate dehydrogenase (GAPDH, a cytosolic marker).

5. Choose the percentage of the gel to be used according to the molecular weight of proteins of interest. In general: 4–5 % gels: >250 kDa; 7.5 % gels: 250–120 kDa; 10 % gels: 120–40 kDa; 13 % gels: 40–15 kDa; 15 % gels: <20 kDa.

6. Apply 5–25 μg total protein of sample to each well of a 0.75–1.0 mm thick gel. For thicker gels (1.5 mm thick), apply up to 25–40 μg in each well.

7. For proteins smaller than 20 kDa, transfer proteins from gel to PVDF membrane at 350 mA for 1 h in transfer buffer. For proteins larger than 120 kDa, transfer to PVDF membrane at 350 mA for 140 min in transfer buffer supplemented with 0.05 % SDS.

8. Some antibodies require the use of BSA (1–5 %) as a blocking agent. Read the vendor's instructions before following the method.

9. All incubation and washing steps are carried out while gently shaking.

10. Alternatively, regular medium without phenol red or Krebs medium can be used, to avoid interferences with the probe's fluorescence.

11. Unspecific binding of the antibodies can be also blocked with 10 % serum from the species in which the secondary antibody was raised.

Acknowledgements

I.V.N. (SFRH/BPD/86534/2012), T.L.S. (SFRH/BPD/75959/2011), T.C.O. (SFRH/BPD/34711/2007) and V.A.S. (SFRH/BPD/31549/2006) are postdoctoral fellows from the Portuguese Foundation for Science and Technology (FCT). Work in the authors laboratory is supported by research grants Pest-C/SAU/LA0001/2013-2014, PTDC/SAU-TOX/117912/2010, PTDC/DTP-FTO/1180/2012 and PTDC/QUI-QUI/101409/2008 funded by Fundação para a Ciência e a Tecnologia (FCT), Portugal, and cofinanced by: "COMPETE—Programa Operacional Factores de Competitividade", QREN and European Union (FEDER—Fundo Europeu de Desenvolvimento Regional).

References

1. Galluzzi L, Kepp O, Trojel-Hansen C et al (2012) Mitochondrial control of cellular life, stress, and death. Circ Res 111:1198–1207

2. Chao DT, Korsmeyer SJ (1998) BCL-2 family: regulators of cell death. Annu Rev Immunol 16:395–419

3. Reed JC, Zha H, Aime-Sempe C et al (1996) Structure-function analysis of Bcl-2 family proteins. Regulators of programmed cell death. Adv Exp Med Biol 406:99–112

4. Scorrano L, Oakes SA, Opferman JT et al (2003) BAX and BAK regulation of endoplasmic reticulum Ca2+: a control point for apoptosis. Science 300:135–139

5. Bernardi P (2013) The mitochondrial permeability transition pore: a mystery solved? Front Physiol 4:95

6. Precht TA, Phelps RA, Linseman DA et al (2005) The permeability transition pore triggers Bax translocation to mitochondria during neuronal apoptosis. Cell Death Differ 12:255–265

7. Liu B, Chen Y, St Clair DK (2008) ROS and p53: a versatile partnership. Free Radic Biol Med 44:1529–1535

8. Vazquez A, Bond EE, Levine AJ et al (2008) The genetics of the p53 pathway, apoptosis and cancer therapy. Nat Rev Drug Discov 7:979–987

9. Galluzzi L, Morselli E, Kepp O et al (2011) Mitochondrial liaisons of p53. Antioxid Redox Signal 15:1691–1714

10. Liu Y, Kulesz-Martin M (2001) p53 protein at the hub of cellular DNA damage response pathways through sequence-specific and non-sequence-specific DNA binding. Carcinogenesis 22:851–860

11. Laptenko O, Prives C (2006) Transcriptional regulation by p53: one protein, many possibilities. Cell Death Differ 13:951–961

12. Yamaguchi H, Woods NT, Piluso LG et al (2009) p53 acetylation is crucial for its transcription-independent proapoptotic functions. J Biol Chem 284:11171–11183

13. Shieh SY, Ikeda M, Taya Y et al (1997) DNA damage-induced phosphorylation of p53 alleviates inhibition by MDM2. Cell 91:325–334

14. Cox ML, Meek DW (2010) Phosphorylation of serine 392 in p53 is a common and integral event during p53 induction by diverse stimuli. Cell Signal 22:564–571

15. Yap DB, Hsieh JK, Zhong S et al (2004) Ser392 phosphorylation regulates the oncogenic

function of mutant p53. Cancer Res 64: 4749–4754

16. Hu W, Feng Z, Levine AJ (2012) The regulation of multiple p53 stress responses is mediated through MDM2. Genes Cancer 3: 199–208

17. Tasdemir E, Maiuri MC, Galluzzi L et al (2008) Regulation of autophagy by cytoplasmic p53. Nat Cell Biol 10:676–687

18. Vega-Naredo I, Caballero B, Sierra V et al (2009) Sexual dimorphism of autophagy in Syrian hamster Harderian gland culminates in a holocrine secretion in female glands. Autophagy 5:1004–1017

19. Yamamoto H, Ozaki T, Nakanishi M et al (2007) Oxidative stress induces p53-dependent apoptosis in hepatoblastoma cell through its nuclear translocation. Genes Cells 12:461–471

20. Salminen A, Kaarniranta K (2009) SIRT1: regulation of longevity via autophagy. Cell Signal 21:1356–1360

21. Chipuk JE, Kuwana T, Bouchier-Hayes L et al (2004) Direct activation of Bax by p53 mediates mitochondrial membrane permeabilization and apoptosis. Science 303:1010–1014

22. Sardao VA, Oliveira PJ, Holy J et al (2009) Doxorubicin-induced mitochondrial dysfunction is secondary to nuclear p53 activation in H9c2 cardiomyoblasts. Cancer Chemother Pharmacol 64:811–827

23. Moll UM, Wolff S, Speidel D et al (2005) Transcription-independent pro-apoptotic functions of p53. Curr Opin Cell Biol 17:631–636

24. Mihara M, Erster S, Zaika A et al (2003) p53 has a direct apoptogenic role at the mitochondria. Mol Cell 11:577–590

25. Zhao Y, Chaiswing L, Velez JM et al (2005) p53 translocation to mitochondria precedes its nuclear translocation and targets mitochondrial oxidative defense protein-manganese superoxide dismutase. Cancer Res 65: 3745–3750

26. Green ML, Pisano MM, Prough RA et al (2013) Release of targeted p53 from the mitochondrion as an early signal during mitochondrial dysfunction. Cell Signal 25: 2383–2390

27. Goldstein JC, Munoz-Pinedo C, Ricci JE et al (2005) Cytochrome c is released in a single step during apoptosis. Cell Death Differ 12: 453–462

28. Martinez-Ruiz G, Maldonado V, Ceballos-Cancino G et al (2008) Role of Smac/DIABLO in cancer progression. J Exp Clin Cancer Res 27:48

29. Kaufmann T, Strasser A, Jost PJ (2012) Fas death receptor signalling: roles of Bid and XIAP. Cell Death Differ 19:42–50

30. Kilbride SM, Prehn JH (2013) Central roles of apoptotic proteins in mitochondrial function. Oncogene 32:2703–2711

31. Susin SA, Lorenzo HK, Zamzami N et al (1999) Molecular characterization of mitochondrial apoptosis-inducing factor. Nature 397:441–446

32. Cande C, Cohen I, Daugas E et al (2002) Apoptosis-inducing factor (AIF): a novel caspase-independent death effector released from mitochondria. Biochimie 84:215–222

33. Lipton SA, Bossy-Wetzel E (2002) Dueling activities of AIF in cell death versus survival: DNA binding and redox activity. Cell 111: 147–150

34. Ye H, Cande C, Stephanou NC et al (2002) DNA binding is required for the apoptogenic action of apoptosis inducing factor. Nat Struct Biol 9:680–684

35. Baritaud M, Boujrad H, Lorenzo HK et al (2010) Histone H2AX: the missing link in AIF-mediated caspase-independent programmed necrosis. Cell Cycle 9:3166–3173

36. Cande C, Vahsen N, Kouranti I et al (2004) AIF and cyclophilin A cooperate in apoptosis-associated chromatinolysis. Oncogene 23: 1514–1521

Chapter 15

Mitophagy and Mitochondrial Balance

Simone Patergnani and Paolo Pinton

Abstract

Mitochondria are highly dynamic organelles, with a morphology ranging from small roundish elements to large interconnected networks. This fine architecture has a significant impact on mitochondrial homeostasis, and mitochondrial morphology is highly connected to specific cellular process. Autophagy is a catabolic process in which cell constituents, including proteins and organelles, are delivered to the lysosomal compartment for degradation. Autophagy has multiple physiological functions and recent advances have demonstrated that this process is linked to different human diseases, such as cancer and neurodegenerative disorders.

In particular, it has been found that autophagy is a key determinant for the life span of mitochondria through a particularly fine-tuned mechanism called mitophagy, a selective form of autophagy, which ensures the preservation of healthy mitochondria through the removal of damaged or superfluous mitochondria. Mitophagy has been found to be altered in several pathologies and aberrant or excessive levels of this process are found in common human disorders. Thus, the measurement of the mitophagy levels is of fundamental relevance to elucidate the molecular mechanism of this process and, most importantly, its role in cellular homeostasis and disease.

In this review, we will provide an overview of the current methods used to measure mitophagic levels, with particular emphasis on the techniques based on fluorescent probes.

Key words Mitochondria, Autophagy, Mitophagy, Fluorescent probes, GFP-LC3

1 Introduction

In the presence of oxygen, cells are able to metabolize glucose by oxidation of glycolytic pyruvate in the mitochondrial tricarboxylic acid (TCA) cycle. At the end of this cycle, the NADH (nicotinamide adenine dinucleotide, reduced) is used by oxidative phosphorylation to boost ATP production. Due to this, it was possible to coin the popular term that identifies mitochondria as the "powerhouse of the cell."

A growing body of recent evidence suggests that mitochondria are part of a more complex cellular signalling network and play a central role in several physiological processes (such as cell proliferation, autophagy, and apoptosis) [1–3], in cellular processes like metabolism [4], during the stress response [5], and in the regulation

Carlos M. Palmeira and Anabela P. Rolo (eds.), *Mitochondrial Regulation*, Methods in Molecular Biology, vol. 1241, DOI 10.1007/978-1-4939-1875-1_15, © Springer Science+Business Media New York 2015

of the homeostasis of second messengers, such as calcium (Ca^{2+}) and reactive oxygen species (ROS) [6, 7].

In short, it is clear that the mitochondrial compartment drives essential processes for a correct cell physiology and cell fate. As demonstration of this, alteration of the normal homeostasis of mitochondria is always correlated to common human pathologies [8, 9].

However, even though mitochondria are critical and indispensable elements, the unchecked existence of mitochondria within the cell could become very dangerous. Aged or damaged mitochondria could induce an excessive ROS production, which leads to several mitochondrial dysfunctions, prompting the release of apoptosis-promoting factors and the consequent damage to neighboring mitochondria. As a result, the well-being of the cell could be deeply undermined.

Based on this knowledge, it is so easy to conceive that cells have developed finely tuned mechanisms to supervise mitochondrial quality and quantity. Mitophagy, a catabolic process for lysosome-dependent degradation, has been recently described as a mechanism for the elimination of damaged and unwanted mitochondria [10].

In the last decade, three distinct mechanisms of mitophagy have been identified: (1) during the transformation from reticulocyte to a mature erythrocyte, all the internal organelles, including mitochondria, are eliminated. This particular form of mitophagy involves a Bcl-2-related mitochondria outer membrane protein Nix (also known as BNIP3L) and the microtubule-associated protein light chain 3 (LC3; also called MAP1LC3 or LC3B, the ortholog of yeast ATG8), the principal autophagosome-associated protein [10]; (2) the second mitophagic mechanism was observed in yeast, where the autophagy-related gene 32 (ATG32) protein, localized on the mitochondrial outer membrane, ensures selective sequestration of mitochondria by the recruitment of the canonical effectors of the autophagic machinery [11]; (3) the third (and also the most studied) pathway for the elimination of damaged mitochondria by mitophagy in mammals is a fine-tuned mechanism mediated by two Parkinson Disease (PD)-associated genes: PINK1 (PTEN-induced putative protein kinase 1) and PARK2/PARKIN [12]. When a subset of mitochondria suffer a collapse of Ψ_m (mitochondrial membrane potential), induced by stressors or uncoupler agents, PINK1 and Parkin cooperate together for the removal of damaged mitochondria [13]. Normally, when the Ψ_m is intact, low levels of the serine/threonine PINK1 are found in mitochondria, because it is constantly imported (probably via the TIM/TOM complex) and cleaved by mitochondrial proteases by the inner membrane presenilin-associated rhomboid-like protease (PARL) and the mitochondrial-processing protease (MPP) [14, 15]. Upon loss of Ψ_m, these mechanisms are affected and PINK1 rapidly accumulates on the outer mitochondrial surface and acts as a marker for

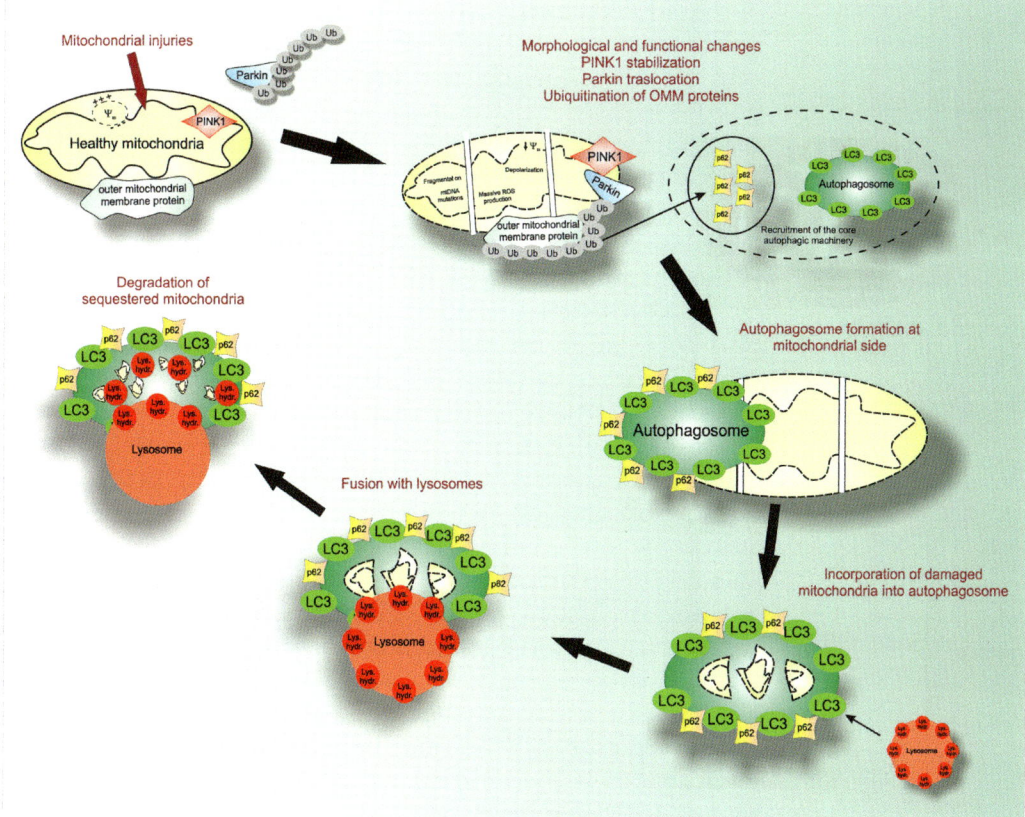

Fig. 1 Mechanism of mitophagy. Following mitochondrial injuries the kinase PINK1 accumulates on the OMM, where it recruits the E3 ubiquitin ligase Parkin to mitochondria. Parkin then promotes the ubiquitination of OMM proteins inducing the recruitment of p62 to clustered mitochondria. Finally, p62 accumulates on mitochondria, binds to parkin-ubiquitylated mitochondrial substrates, mediates clumping of mitochondria and links ubiquitinated substrates to LC3 to facilitate the autophagic degradation of ubiquitinated proteins. *Ub* ubiquitin, Ψ_m mitochondrial membrane potential, *OMM* outer mitochondrial membrane, *LC3* microtubule-associated protein light chain 3, *Lys.hydr* lysosomal hydrolase, *PINK1* PTEN-induced putative protein kinase 1

mitochondrial damage. As a consequence, PINK1 leads to the recruitment of Parkin from the cytosol to mitochondria, where it mediates the ubiquitination of numerous outer mitochondrial membrane proteins [16]. In this way, the docking site for the Ub (ubiquitin)-binding adaptor SQSTM1/p62 is created, which then accumulates in mitochondria and facilitates the recruitment of damaged mitochondria to autophagosome by binding the LC3-interacting region (LIR) motif of LC3 [17] (Fig. 1).

1.1 Methods to Monitor Mitophagy

As outlined above, the removal of damaged mitochondria is a critical aspect for the well-being of cells. Alterations of mitophagy pathways are increasingly recognized in a number of human diseases, including cancer and neurodegeneration [18]. To better understand the role and mechanism of mitophagy in these settings, in the last decade several methods are been developed to monitor and visualize this catabolic process (*see* **Notes 1** and **2**). Like conventional autophagy, also for mitochondrial autophagy it is possible to assess the incorporation of "wasted" mitochondria into the autophagosome by electron microscopy or the release of radioactively labelled cellular proteins [19, 20]. These techniques provide confirmation of mitochondrial autophagy or clearance, but present some caveats and limitations when it comes to the quantification of mitophagy.

Another useful method to quantify the loss of the mitochondrial pool due to the mitophagic process is the measurement of expression levels of mitochondrial proteins employing immunoblot assays [21]. Since protein synthesis and protein degradation are critical and typical aspects of the mitochondrial pools of protein, and also considering that intermembrane space proteins are frequently lost after permeability transition, it is recommended to analyze the total cellular levels of proteins linked to different mitochondrial subcompartments. Furthermore, to ensure that the protein loss is limited to the mitochondrial compartment, it is suggested to perform an immunoblot analysis against non mitochondrial protein (e.g., endoplasmic reticulum proteins).

A robust way to investigate mitophagy is through the use of fluorescent probes, which may be used to visualize sequestered mitochondria in the autophagosome and the subsequent delivery to the lysosomal compartment [22].

Below, we describe the most used methods to quantify the mitophagic process by fluorescent microscopy and we present a detailed protocol using fluorescent probes to evaluate the selective removal of damaged mitochondria by mitophagy.

1.2 Monitoring Mitophagy Using Fluorescent Probes

At the moment, the main methods to measure mitophagic activity using fluorescent microscopy are based on the simultaneous visualization of mitochondria and autophagosomes with the autophagosome-specific marker LC3. Like autophagy, mitophagy is a dynamic and multistep process that can be modulated at different steps. Based on these observations, researchers have developed novel fluorescent-based techniques to monitor the activity of the mitophagic machinery. Some of them are designed to monitor the delivery of mitochondria to the lysosome. To visualize the amount of fused organelles, mitochondria may be labelled with a mitochondrial marker without significant membrane potential dependence (MitoTracker® dyes) (*see* **Note 3**) and the lysosome may be stained with fluorescent probes (such as the lysosomotropic LysoTracker® dyes) (*see* **Note 4**) or a lysosomal marker (e.g., LAMP1-GFP) [23, 24]. A similar approach, to evaluate the incorporation of mitochondria into the autolysosome makes use of a mitochondria-targeted version of the pH-dependent Keima protein, named mito-Keima. When mitochondria are sequestered into the lysosomal compartment, the peak of excitation of this modified protein shifts from 440 nm (high pH) to 586 nm (acidic pH) [25].

One key role of mitophagy is the removal of dysfunctional and aged mitochondria so as to maintain mitochondrial turnover and cellular homeostasis. Thus, monitoring these mitochondrial changes is of great importance. A novel fluorescent tool, named "MitoTimer," is a mitochondria-targeted mutant of the red fluorescent protein Fluorescent Timer, known as DsRed1-E5, which shows a fluorescence shift from green to red as the protein matures. As such, this mitochondrial fluorescent protein may be a very useful method to visualize real-time mitochondrial exchange in living cells [26].

Other methods are based on the peculiarity that MAP1LC3 is not the only marker for autophagy, and several other proteins are also involved in this catabolic process. Importantly, a number of these markers are available to perform fluorescent imaging. As reported above, PINK1 and Parkin associate together to label damaged mitochondria, which will be marked for selective degradation via autophagy. In particular, a key aspect for the execution of mitophagy is the translocation of Parkin from cytosol to mitochondria, an event that can be recognized through the use of fluorescent microscopy based on the simultaneous visualization of mitochondria (using mitochondrial fluorescent probes, such as MitoTracker Green, or a mitochondrial marker, such as mitochondria-targeted GFP, mtGFP) and the Parkin protein (using fluorescent recombinant chimeras, such as mCherry-Parkin) [27]. Another light-based method to recognize the specific initiation of Parkin-mediated mitophagy uses a genetically encoded mitochondria-matrix targeting photosensitizer "KillerRed-dMito" and the fluorescent recombinant Parkin-chimera YFP-Parkin. KillerRed is a photosensitizer which produces ROS when illuminated with 599-nm light; as result, mitochondria in the selectively illuminated area become impaired. This mitochondrial damage is recognized by Parkin which translocates to the mitochondrial surface and induces mitochondrial clearance by the LC3-coated autophagic structures [28]. Thanks to this new method, it will be possible to monitor the Parkin-mediated mitophagy in a temporally controlled fashion and it could be useful to identify the high-tuned (but still unknown) molecular mechanism in Parkin-mediated mitophagy (*see* **Note 5**).

Even though different methods and several markers for selective autophagy are been unveiled, the final step of the mitophagic process always implies the incorporation of damaged mitochondria in LC3-coated vesicles (*see* **Note 6**). To date, the simultaneous visualization of mitochondria and autophagosome remains of fundamental importance for a correct mitophagic analysis [21, 29].

In next section, we describe in detail the direct method to provide confirmation of the incorporation of damaged mitochondria into autophagosome by fluorescent microscopy using a mitochondria-targeted RFP (mtDsRed) in combination with the autophagosomal marker GFP-MAP1LC3B (GFP-LC3).

2   Materials

2.1   Cell Culture and Reagents

1. Cell lines: Rst oligodendrocyte precursor cells (OPCs).

2. Culture medium for OPCs: Dulbecco's modified Eagle's medium (DMEM), 4 mM l-glutamine, 1 mM sodium pyruvate, 0.1 % bovine serum albumin (BSA), 50 mg/ml Apo-transferrin, 5 mg/ml insulin, 30 nM sodium selenite, 10 nM d-biotin, 10 nM hydrocortisone, 100 U/ml penicillin, 100 mg/ml streptomycin.

3. 10 μg/ml PDGF (platelet-derived growth factor)-AA: stock prepared in distilled water or media, filtered and stored at −20 °C; use at 10 ng/ml.

4. 10 μg/ml bFGF (basic fibroblast growth factor): stock prepared in distilled water or media, filtered and stored at −20 °C; use at 10 ng/ml.

5. 10 μM FCCP (Carbonyl cyanide 4-trifluoromethoxyphenyl hydrazone), dissolved in ethanol and stored at −20 °C; use at 100 nM for OPCs.

6. Lipofectamine 2000 Reagent (Invitrogen, Life Technologies, Carlsbad, CA).

7. Glass coverslips (24 mm in diameter).

8. Poly-L-Ornithine (10 mg/ml stock); use at 100 μg/ml.

9. Six-well culture plates.

2.2   Image (Fluorescence)-Based Analysis of Mitophagy

1. Inverted Nikon LiveScan Swept Field Confocal Microscope (SFC) Eclipse Ti equipped with NIS-Elements microscope imaging software, an Andor DU885 electron multiplying charge-coupled device (EM-CCD) camera (Andor Technology Ltd, Belfast, Northern Ireland) and a CFI Plan Apo VC60XH objective (numerical aperture, 1.4) (Nikon Instruments, Melville, NY).

2. Autophagosomal marker GFP-MAP1LC3B (GFP-LC3). For further detailed technical descriptions about this plasmid, please refer to ref. [3].

3. Mitochondria-targeted RFP (mtDsRed). For further detailed technical descriptions about this plasmid, please refer to ref. [3].

3   Methods

3.1   Measuring Mitophagy with GFP-LC3 and mtDsRed

In order to obtain an optimal measurement of the degree of colocalization, it is necessary to analyze high-quality images captured at high magnification (60× or 100×). We use an inverted Nikon LiveScan Microscope (SFC) Eclipse Ti equipped with NIS-Elements microscope imaging software, an Andor DU885 electron multiplying charge-coupled device (EM-CCD) camera and a CFI Plan Apo VC60XH objective (numerical aperture, 1.4). To improve resolution and the quality of images acquired, we use a 1.5× amplifier.

3.2 Sample Preparation and Transfection

Cells are seeded on glass coverslips (24 mm in diameter) and allowed to grow until 50 % confluence. After seeding the cells, wait for at least 24 h. The cells are then transfected using appropriate transfection methods (Ca^{2+}-phosphate, lipoamines, electroporation, or adenoviral vectors) with a mitochondria-targeted RFP (mtDsRed) in combination with the autophagosomal marker GFP-MAP1LC3B (GFP-LC3).

Conventionally cancerous and immortalized cell lines may be easily transfect with Ca^{2+}-phosphate or lipoamines transfection methods. Primary cultures, which are notably "hard-to-transfect cells," request electroporation, lipoamines or adenoviral vectors. After transfection, wait for 48 h and perform live fluorescence imaging (*see* **Note 7**). In alternative, it is possible to fix cells with a 2 % formaldehyde solution (*see* **Note 8**).

As certain cell lines possess too low levels of organelle clearance, it may useful to pretreat cells with the potassium ionophore valinomycin or with uncoupling agents [as like CCCP (Carbonyl cyanide m-chlorophenylhydrazone) or FCCP (Carbonyl cyanide 4-trifluoromethoxyphenyl hydrazone)] to recognize their mitophagic activity [30]. These chemical compounds provoke the dissipation of Ψ_m, accumulation of PINK1 on depolarized mitochondria and, finally, Parkin translocation to the mitochondrial outer membrane [31]. Consequently, depolarization of mitochondria by protonophores is highly associated with colocalization of GFP-LC3 on RFP-labelled mitochondria. Another important aspect to keep in mind is that maturation and lysosomal fusion is a very rapid process and the assessment of mitochondrial compartment into autophagosome may be arduous. Treatments with lysosomotropic agents (such as chloroquine, bafilomycin A1, and ammonium chloride) or with lysosomal protease inhibitors (e.g., E64d or pepstatin A), which block the degradation of LC3 with the consequent accumulation of autophagosomes, may facilitate the identification of the cargo of the autophagic vesicles (*see* **Note 9**) [32].

3.3 Sample Preparation and Transfection of Oligodendrocytes Cells

Oligodendrocyte cells were obtained following the protocol developed by Chen et al. [33] and maintained in culture medium for OPCs supplemented with 10 ng/ml PDGF-AA and 10 ng/ml bFGF for a week. The OPCs isolated can be studied in their undifferentiated state or induced to differentiate into immature oligodendroblasts and then into mature oligodendrocytes, replacing PDGF-AA and bFGF with triiodothyronine .

At ~7 days, OPCs obtained were plated at a density of $2 \times 10^4/$ cm^2 on glass coverslips (24 mm in diameter) in six-well culture plates. Cover glasses were previously coated with 100 μg/ml per

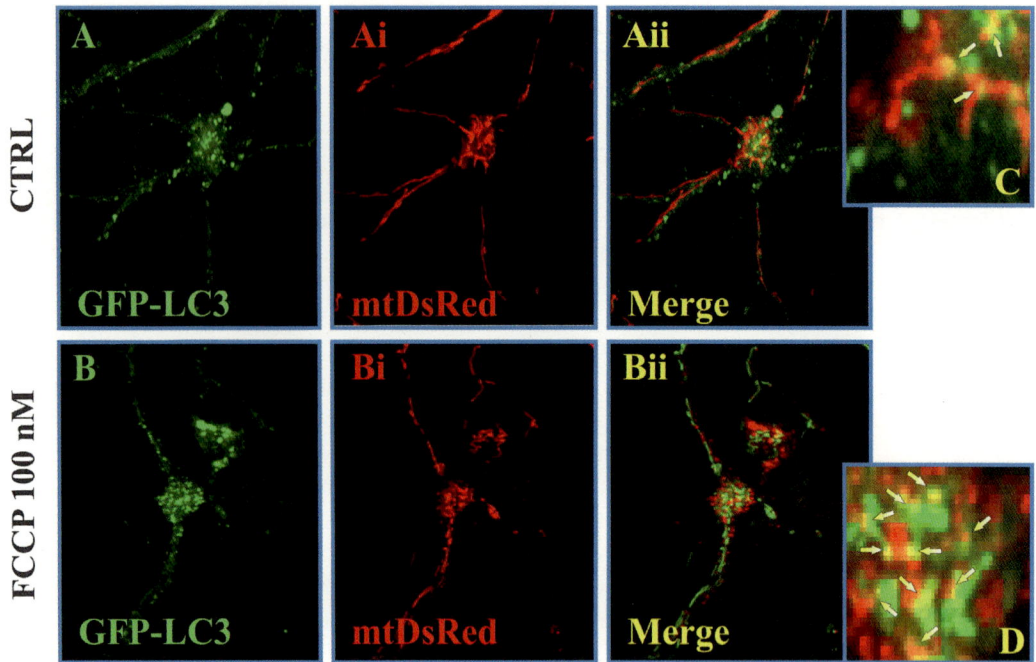

Fig. 2 Analysis of mitophagic sequestration by dual fluorescence. OPCs were transfected with GFP-LC3 and a mitochondrial-targeted red fluorescent protein. Forty-eight hours after transduction, cells were treated with vehicle (*CTRL*) or 100 nM FCCP for 1 h. Confocal images show a significant increase in GFP-LC3 puncta that colocalize with mitochondria in FCCP-treated cells (*arrows*). The images visualize in *blue boxes* (**c** and **d**) are enlargements in **aii** and **bii**, respectively. Cells were imaged on a Nikon LiveScan SFC Microscope Eclipse Ti equipped with VC60XH oil immersion objective with a 1.5× amplifier and appropriate filter sets. Scale bar: 10 μm (**a–aii** and **b–bii**) and 2 μm (**c** and **d**)

well of poly-D-ornithine as follow. Add sufficient quantity of 1× coating solution to glasses and incubate for 1–2 h at 37 °C. Remove solution, wash two times with distilled water and air-dry in a tissue culture hood.

After 24 h from seeding, cells were transfected with 1 μg of plasmid DNA for well (0.5 μg mtDsRed + 0.5 μg GFP-LC3), using the appropriate transfection method (Lipofectamine 2000 reagent). After 4–6 h the transfection medium were replaced with warm medium culture medium for OPCs supplemented with 10 ng/ml PDGF-AA and 10 ng/ml bFGF, and after 48 h the cells were imaged as outlined in Subheading 3.4. Prior to imaging, parallel oligodendrocytes cultures transfected as reported above were treated with the uncoupler agent FCCP 100 nM for 1 h to induce the sequestration of mitochondria into autophagosomes (Fig. 2).

3.4 Measurements

Coverslips (24-mm in diameter) were placed in a stage incubator (Okolab) of the inverted Nikon LiveScan SFC Microscope Eclipse Ti equipped with VC60XH oil immersion objective. Experiments were carried out in basal chemically defined medium supplemented with 10 ng/ml PDGF-AA and 10 ng/ml bFGF. Excitation at

488 nm used a solid state laser (Spectra-Physics, Newport) and fluorescence emission was measured through a 520/30 filter. Laser excitation at 546 nm used a solid state laser (Spectra-Physics, Newport) and fluorescence emission was measured using a 600/20 filter. Laser excitation was attenuated 100- to 1,000-fold to minimize photobleaching and photodamage. In order to obtain statistical significance for measuring the degree of colocalization, we recommend to acquire at least 25–30 images per condition.

3.5   Data Handling/ Processing

Following the experiment, the images acquired can be analyzed directly on the microscope imaging software or by the use of open-source software programs developed to help the interpretation of multidimensional images, such as ImageJ, the Open Microscopy Environment, or VisBio.

Preprocessing of images. Fields of images acquired are not illuminated in a homogeneous fashion; thus, it is recommended to correct for uneven illumination. This image processing may be achieved by correcting the image of the sample using a bright image of an empty field. The correction may be performed using the microscope imaging software or, alternatively, with open-source software programs.

Visualizing colocalization: Conventionally, the most used method to visualize colocalization is the simple merge of the different channels. In our case, GFP-LC3 (green) and mtDsRed (red) give rise to yellow hotspots where the two molecules of interest are present in the same pixels. Thus, to reach a very fast and easy quantification of colocalization, it is possible to perform the simple count of yellow dots and identifying the number of mitochondria sequestered into the autophagosome (Fig. 3a). Alternatively, it is possible to evaluate the intensity profile for the red and green channel of a region of interest, which potentially includes an autophagosome (Fig. 3c–ci).

The simple overlay of the channels of interest possesses, however, some limits. For example, the presence of merged staining (yellow in our case) is highly dependent on the signal intensity of each single channel: to have a reliable colocalization, both images should have similar grey level dynamics. Many software tools and algorithms have been developed to try to address this issue. In particular, to evaluate colocalization events, algorithms performing the so-called Intensity Correlation Coefficient-Based analyses (ICCB; such as Pearson's or Manders' coefficients) and object-based approaches (such as the centroid or intensity center calculation) have been engineered. Recently, a simple public domain tool has been developed for the open-source software ImageJ, named JACoP, which integrates current global statistic methods to evaluate the degree of the colocalization of two channels (Fig. 3b–bi). Using a novel object-based approach, this tool is capable of processing single and composite images and enables an automated colocalization analysis [34].

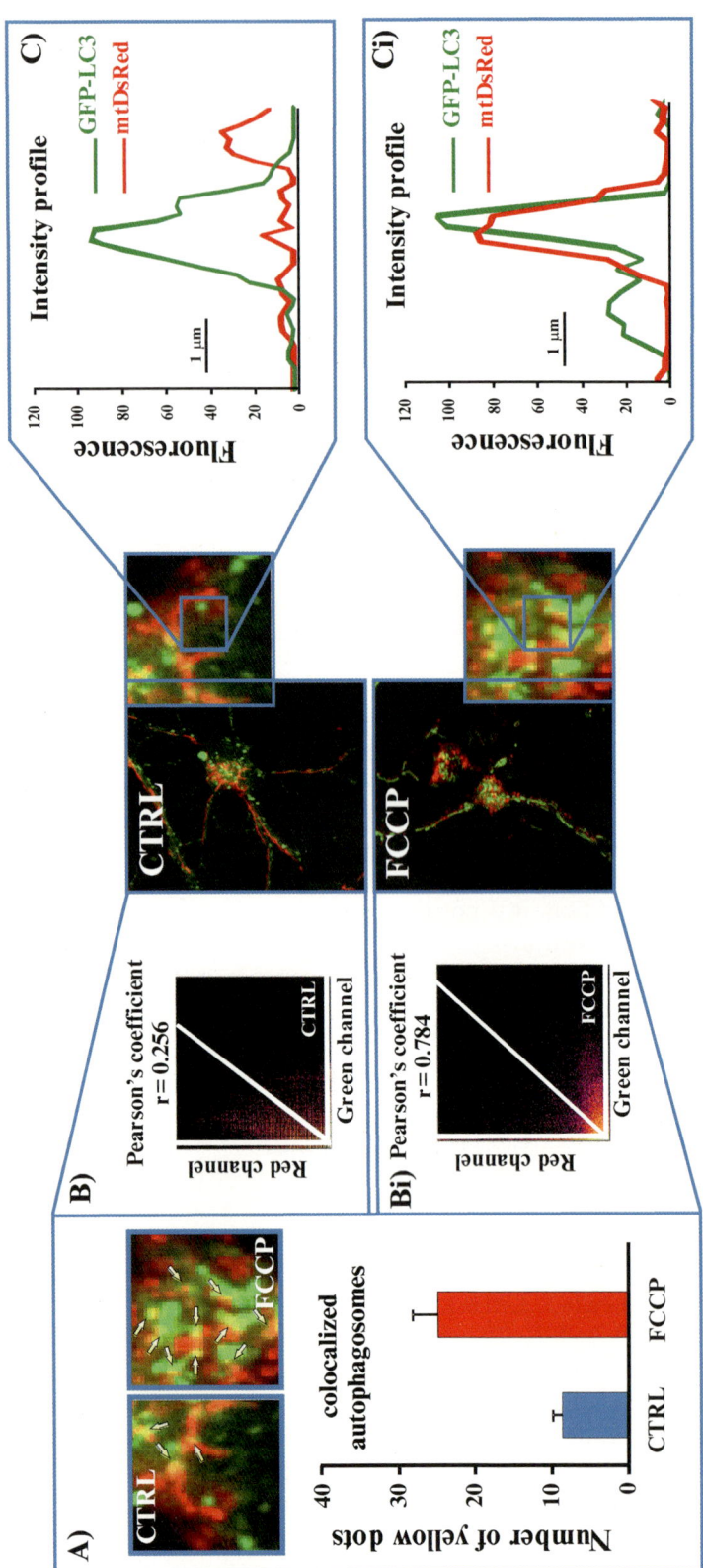

Fig. 3 Methods to quantify colocalization events. (**a**) The most used method to visualize colocalization is the merge of the *red channel* (mitochondria) with the *green one* (autophagosome). As result, it is possible to determine the number of *yellow hotspots* (*arrows*) by manual count or using microscope imaging or open-source software programs. (**b**) Alternatively, it is possible to use Pearson's correlation coefficient as a measure of colocalization of RFP signals with GFP signals. Correlation plot (**b–bi**) is reported corresponding to the left of images. (**c–ci**) Intensity profile of a selected region represents a powerful analytical tool to examine the presence of a possible colocalization. Cells were left for 48 h after plasmids transfection, treated with FCCP 100 nM (*FCCP*) or vehicle (*CTRL*) and imaging. Scale bar: 10 μm (*CTRL* and *FCCP*) and 2 μm (*small blue boxes*)

4   Notes

1. Due to the dynamic nature of mitophagy and the multiple potential factors which regulate this process, it is suggested to perform multiple techniques (rather than a single technique) to evaluate the autophagic removal of damaged mitochondria. For example, immunoblot analysis of the mitochondrial proteins levels is useful to validate data from microscopy studies.

2. It is important to keep in mind that mitochondrial turnover could be promoted by other degradation systems (such as proteosomal degradation) and that several mechanisms are responsible for the disappearance of mitochondrial markers. Based on these observations, it is suggested to evaluate the mitochondrial clearance also with inhibitors of the other major degradation systems.

3. The retention and accumulation of the fluorescent probes, especially for MitoTracker dyes, depend on cell type. It is recommended to determine empirically their optimal concentration in a given cell type under the experimental conditions.

4. Because lysosomotropic agents disrupt the lysosomal acidic pH, these compounds cannot be used with LysoTracker fluorescent probes.

5. Defective or excessive mitophagy is frequently found in several human diseases. Consequently, a novel and accessible method capable of recognizing the alteration of mitophagy and, at the same time, the rapid screening of hypothetical compounds is necessary. Coupling the reported methods with high-content microscopy may be a solution. In this way, it will possible to screen large compound libraries to identify small-molecule modulators of mitophagy. As a demonstration of this theoretical approach, the first mitophagic study performed using high-throughput screening has been recently published. In this work, Hasson and colleagues characterized specific regulators of the PINK1-Parkin-mediated mitophagy. To attain their goal, cells stably expressing GFP-Parkin and a mitochondria-targeted red fluorescent protein were used. The cells were transfected with siRNA duplex in 384-well plates, treated with a chemical mitochondrial uncoupler and the translocation of Parkin to the mitochondrial surface was evaluated by high-content microscopy and automated image analysis [35].

6. In addition to colocalization between mitochondrial labels and markers for autophagy, it may be useful to perform studies in order to quantify the morphological and functional changes associated to the mitophagic removal of mitochondria. To assess these aims, it is possible to evaluate changes in mitochondrial structure (it should be noted that mitochondrial

fragmentation precedes mitophagy) through the employment of MitoTracker, whereas the loss of Ψ_m, a common trigger for mitophagy, can be measured by the use of potentiometric probes (such as JC-1 and TMRM) [36].

7. It is important to consider that certain plasmid DNA and transfection methods may modify the levels of autophagy. To start, it is important to use contaminant-free plasmid DNA. Next, to avoid negative effects induced by transfection, it is suggested to leave the cells for at least 48 h after the transfection. The use of stable cell lines could solve this concern, but: (a) stable cell line are generated by immortalization, a process that modifies markedly the normal physiology of the cell, and (b) it is not always possible or easy to obtain a stable cell line of certain cell types (i.e., nerve cells).

8. The fixation procedure may produce autofluorescent puncta or a reduction for GFP-LC3 staining. To avoid artifacts, it is suggested to compare results of live imaging with imaging after fixation.

9. The optimal lysosomotropic agents or lysosomal protease inhibitors concentration and timetable treatment are highly linked to cell-type. Before perform experiments, one should search for the ideal concentration and time course in order to prevent cytotoxic effects and avoid saturation of the observed LC3-II accumulation.

Acknowledgements

This study was supported by: the Italian Association for Cancer Research (AIRC); Telethon (GGP11139B); local funds from the University of Ferrara; the Italian Ministry of Education, University and Research (COFIN, FIRB, and Futuro in Ricerca); and the Italian Ministry of Health to Paolo Pinton. Simone Patergnani was supported by a FISM (Fondazione Italiana Sclerosi Multipla) research fellowship (2012/B/11).

References

1. Haigis MC, Deng CX, Finley LW, Kim HS, Gius D (2012) SIRT3 is a mitochondrial tumor suppressor: a scientific tale that connects aberrant cellular ROS, the Warburg effect, and carcinogenesis. Cancer Res 72:2468–2472

2. Bonora M, Bononi A, De Marchi E, Giorgi C, Lebiedzinska M, Marchi S, Patergnani S, Rimessi A, Suski JM, Wojtala A et al (2013) Role of the c subunit of the FO ATP synthase in mitochondrial permeability transition. Cell Cycle 12:674–683

3. Patergnani S, Marchi S, Rimessi A, Bonora M, Giorgi C, Mehta KD, Pinton P (2013) PRKCB/protein kinase C, beta and the mitochondrial axis as key regulators of autophagy. Autophagy 9:1367–1385

4. Houtkooper RH, Pirinen E, Auwerx J (2012) Sirtuins as regulators of metabolism and healthspan. Nat Rev Mol Cell Biol 13: 225–238

5. Manoli I, Alesci S, Blackman MR, Su YA, Rennert OM, Chrousos GP (2007) Mitochondria as key

components of the stress response. Trends Endocrinol Metab 18:190–198

6. Giorgi C, Baldassari F, Bononi A, Bonora M, De Marchi E, Marchi S, Missiroli S, Patergnani S, Rimessi A, Suski JM et al (2012) Mitochondrial Ca(2+) and apoptosis. Cell Calcium 52:36–43

7. Marchi S, Giorgi C, Suski JM, Agnoletto C, Bononi A, Bonora M, De Marchi E, Missiroli S, Patergnani S, Poletti F et al (2012) Mitochondria-ros crosstalk in the control of cell death and aging. J Signal Transduct 2012:329635

8. Wallace DC (1999) Mitochondrial diseases in man and mouse. Science 283:1482–1488

9. Itoh K, Nakamura K, Iijima M, Sesaki H (2013) Mitochondrial dynamics in neurodegeneration. Trends Cell Biol 23:64–71

10. Youle RJ, Narendra DP (2011) Mechanisms of mitophagy. Nat Rev Mol Cell Biol 12:9–14

11. Kanki T, Wang K, Cao Y, Baba M, Klionsky DJ (2009) Atg32 is a mitochondrial protein that confers selectivity during mitophagy. Dev Cell 17:98–109

12. Springer W, Kahle PJ (2011) Regulation of PINK1-Parkin-mediated mitophagy. Autophagy 7:266–278

13. Narendra D, Tanaka A, Suen DF, Youle RJ (2008) Parkin is recruited selectively to impaired mitochondria and promotes their autophagy. J Cell Biol 183:795–803

14. Meissner C, Lorenz H, Weihofen A, Selkoe DJ, Lemberg MK (2011) The mitochondrial intramembrane protease PARL cleaves human Pink1 to regulate Pink1 trafficking. J Neurochem 117:856–867

15. Narendra DP, Jin SM, Tanaka A, Suen DF, Gautier CA, Shen J, Cookson MR, Youle RJ (2010) PINK1 is selectively stabilized on impaired mitochondria to activate Parkin. PLoS Biol 8:e1000298

16. Jin SM, Lazarou M, Wang C, Kane LA, Narendra DP, Youle RJ (2010) Mitochondrial membrane potential regulates PINK1 import and proteolytic destabilization by PARL. J Cell Biol 191:933–942

17. Jin SM, Youle RJ (2012) PINK1- and Parkin-mediated mitophagy at a glance. J Cell Sci 125:795–799

18. Lu H, Li G, Liu L, Feng L, Wang X, Jin H (2013) Regulation and function of mitophagy in development and cancer. Autophagy 9:1720–1736

19. Dagda RK, Cherra SJ III, Kulich SM, Tandon A, Park D, Chu CT (2009) Loss of PINK1 function promotes mitophagy through effects on oxidative stress and mitochondrial fission. J Biol Chem 284:13843–13855

20. Frank M, Duvezin-Caubet S, Koob S, Occhipinti A, Jagasia R, Petcherski A, Ruonala MO, Priault M, Salin B, Reichert AS (2012) Mitophagy is triggered by mild oxidative stress in a mitochondrial fission dependent manner. Biochim Biophys Acta 1823:2297–2310

21. Klionsky DJ, Abdalla FC, Abeliovich H, Abraham RT, Acevedo-Arozena A, Adeli K, Agholme L, Agnello M, Agostinis P, Aguirre-Ghiso JA et al (2012) Guidelines for the use and interpretation of assays for monitoring autophagy. Autophagy 8:445–544

22. Dolman NJ, Chambers KM, Mandavilli B, Batchelor RH, Janes MS (2013) Tools and techniques to measure mitophagy using fluorescence microscopy. Autophagy 9:1653–1662

23. Rodriguez-Enriquez S, Kim I, Currin RT, Lemasters JJ (2006) Tracker dyes to probe mitochondrial autophagy (mitophagy) in rat hepatocytes. Autophagy 2:39–46

24. Pankiv S, Clausen TH, Lamark T, Brech A, Bruun JA, Outzen H, Overvatn A, Bjorkoy G, Johansen T (2007) p62/SQSTM1 binds directly to Atg8/LC3 to facilitate degradation of ubiquitinated protein aggregates by autophagy. J Biol Chem 282:24131–24145

25. Katayama H, Kogure T, Mizushima N, Yoshimori T, Miyawaki A (2011) A sensitive and quantitative technique for detecting autophagic events based on lysosomal delivery. Chem Biol 18:1042–1052

26. Hernandez G, Thornton C, Stotland A, Lui D, Sin J, Ramil J, Magee N, Andres A, Quarato G, Carreira RS et al (2013) MitoTimer: a novel tool for monitoring mitochondrial turnover. Autophagy 9:1852–1861

27. Matsuda N, Sato S, Shiba K, Okatsu K, Saisho K, Gautier CA, Sou YS, Saiki S, Kawajiri S, Sato F et al (2010) PINK1 stabilized by mitochondrial depolarization recruits Parkin to damaged mitochondria and activates latent Parkin for mitophagy. J Cell Biol 189:211–221

28. Yang JY, Yang WY (2011) Spatiotemporally controlled initiation of Parkin-mediated mitophagy within single cells. Autophagy 7:1230–1238

29. Zhu J, Dagda RK, Chu CT (2011) Monitoring mitophagy in neuronal cell cultures. Methods Mol Biol 793:325–339

30. Ashrafi G, Schwarz TL (2013) The pathways of mitophagy for quality control and clearance of mitochondria. Cell Death Differ 20.31 12

31. Rakovic A, Shurkewitsch K, Seibler P, Grunewald A, Zanon A, Hagenah J, Krainc D, Klein C (2013) Phosphatase and tensin homolog

(PTEN)-induced putative kinase 1 (PINK1)-dependent ubiquitination of endogenous Parkin attenuates mitophagy: study in human primary fibroblasts and induced pluripotent stem cell-derived neurons. J Biol Chem 288:2223–2237

32. Mizushima N, Yoshimori T, Levine B (2010) Methods in mammalian autophagy research. Cell 140:313–326

33. Chen Y, Balasubramaniyan V, Peng J, Hurlock EC, Tallquist M, Li J, Lu QR (2007) Isolation and culture of rat and mouse oligodendrocyte precursor cells. Nat Protoc 2:1044–1051

34. Bolte S, Cordelieres FP (2006) A guided tour into subcellular colocalization analysis in light microscopy. J Microsc 224:213–232

35. Hasson SA, Kane LA, Yamano K, Huang CH, Sliter DA, Buehler E, Wang C, Heman-Ackah SM, Hessa T, Guha R et al (2013) High-content genome-wide RNAi screens identify regulators of parkin upstream of mitophagy. Nature 504(7479):291–295

36. Duchen MR, Surin A, Jacobson J (2003) Imaging mitochondrial function in intact cells. Methods Enzymol 361:353–389

INDEX

Carlos M. Palmeira and Anabela P. Rolo (eds.), *Mitochondrial Regulation*, Methods in Molecular Biology,
vol. 1241, DOI 10.1007/978-1-4939-1875-1, © Springer Science+Business Media New York 2015